THE PHYSICS COMPANION

A C Fischer-Cripps

I₀P

Institute of Physics Publishing
Bristol and Philadelphia

British Library Cataloguing-in-Publication Data

A catalogue record for this book is available from the British Library.

ISBN 0 7503 0953 9

Library of Congress Cataloging-in-Publication Data are available

Commissioning Editor: Tom Spicer
Production Editor: Simon Laurenson
Production Control: Sarah Plenty
Cover Design: Frédérique Swist
Marketing: Nicola Newey and Verity Cooke

Published by Institute of Physics Publishing, wholly owned by The Institute of Physics, London

Institute of Physics Publishing, Dirac House, Temple Back, Bristol BS1 6BE, UK

US Office: Institute of Physics Publishing, The Public Ledger Building, Suite 929, 150 South Independence Mall West, Philadelphia, PA 19106, USA

Printed in the UK by MPG Books Ltd, Bodmin, Cornwall

This book is dedicated to Rod Cameron, OSA, my former high school science teacher, who, when I asked him "Why do we study these things?" replied "To know the truth", thus beginning my career as a physicist.

Contents

Preface

When I was a physics student, I found things hard going. My physics text book and my professor assumed that I knew far too much, and as a result, I did not do very well in examinations despite my personal interest in the subject. Later, when it was my turn to be in front of the class, I decided to write up my lecture notes in such a way so as to provide clear and succinct accounts of the wide variety of topics to be found in first year university physics. This book is the result, and I hope that you will find it helpful for your understanding of what is an extraordinary subject.

In writing this book, I was assisted and encouraged by Jim Franklin, Hillary Goldsmith, Suzanne Hogg, Walter Kalceff, Les Kirkup, John O'Connor, Joe Wolfe, Tom von Foerster, my colleagues at the University of Technology, Sydney, and all my former students.

Special thanks are due to Dr. Robert Cheary and Prof. Richard Collins. I hope that through this book you will in turn benefit from whatever I have been able to transmit of their professional and enthusiastic approach to physics that was my privilege to experience as their student. My sincere thanks to my wife and family for their unending encouragement and support. Finally, I thank Tom Spicer and the editorial and production team at Institute of Physics Publishing for their very professional and helpful approach to the whole publication process.

Tony Fischer-Cripps,
Killarney Heights, Australia, 2003

Part 1

Thermal Physics

1.1 Temperature

Summary

Thermal equilibrium:

Two systems are in thermal equilibrium when they are at the same temperature.

Zeroth law of thermodynamics:

When two systems are in thermal equilibrium with a third, they are also in thermal equilibrium with each other.

Temperature scales:

Absolute zero: 0K $-273.15\,^{\circ}\text{C}$
Ice point: 273.15K $0\,^{\circ}\text{C}$
Triple point: 273.16K $0.01\,^{\circ}\text{C}$
Boiling point   373.15K $100\,^{\circ}\text{C}$

$$^{\circ}\text{C} = \left(^{\circ}\text{F} - 32\right)\frac{100}{180}$$

1.1.1 Thermodynamic systems

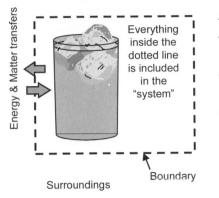

Everything inside the dotted line is included in the "system"

Boundary

Surroundings

Energy & Matter transfers

A **thermodynamic system** is:

- an isolated portion of space or quantity of matter.

- separated from the surroundings by a boundary.

- analysed by considering transfers of energy and matter across the boundary between the system and the surroundings.

Describing a thermodynamic system:

- macroscopic quantities ———→
- microscopic quantities
  ↓
  Kinetic energy of molecules.

- Volume
- Pressure
- Temperature
- Mass

Consider two separate thermodynamic systems A and B which are initially at different temperatures and are now brought into contact.

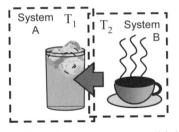

System A T_1 T_2 System B

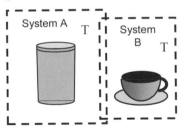

System A T System B T

Heat flows from the system at high temperature to the system at low temperature.

Both systems eventually reach the same **temperature**.

At this condition, System A is now in **thermal equilibrium** with System B.

When two systems are separately in thermal equilibrium with a third, they are also in **thermal equilibrium** with each other. This statement is known as the **zeroth law of thermodynamics**.

1.1.2 Temperature

What is **temperature?** The microscopic answer is that temperature is a measure of the total **kinetic energy** of the molecules in a system.

How can it be measured?
• Our senses (touch, sight)
• With a thermometer

The measurement of temperature is naturally associated with the definition of a **temperature scale**.

Celsius temperature scale:
defined such that
0 °C = ice point of water
100 °C = boiling point of water.

Fahrenheit temperature scale:
defined such that
32 °F = ice point of water
212 °F = boiling point of water

Note: **Standard atmospheric pressure** (1 atm) is defined as 760 mm Hg (ρ = 13.5951 g cm^{-3}) at g = 9.80665 ms^{-2}.

$$^{\circ}C = \left(^{\circ}F - 32\right)\frac{100}{180}$$

The **International Temperature Scale** is based on the definition of a number of basic **fixed points**. The fixed points cover the range of temperatures to be normally found in industrial processes. The most commonly used fixed points are:

1. Temperature of equilibrium between ice and air-saturated water at normal atmospheric pressure (ice point) is: 0.000 °C

2. Temperature of equilibrium between liquid water and its vapour at a pressure of 1 atm pressure (steam point) is: 100.000 °C

The ice and steam points are convenient fixed points. A more reproducible fixed point is the **triple point** of water.

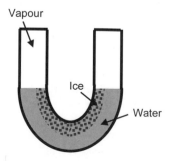

Vapour

Ice

Water

The state of pure water existing as an equilibrium mixture of ice, liquid and vapour. Let the temperature of water at its triple point be equal to 273.16K. This assignment corresponds to a an ice point of 273.15K or 0°C. *Triple point is used as the standard fixed point because it is reproducible.*

1.1.3 Kelvin temperature scale

A **constant volume gas thermometer** (**CVGT**) is a special
thermometer that gives a reading of pressure as an indication
of temperature.

Ideal gas ←
- Molecules occupy a
  very small volume
- Molecules are at
  relatively large
  distances from each
  other
- Collisions between
  molecules are elastic

- Gas in chamber is at low pressure
  (so gas is "ideal")
- Gas must not condense into a liquid
  (Helium is a good gas to use for
  low temperatures)
- Pressure of the gas is an indication
  of temperature
- Must maintain constant gas volume

kPa

The CVGT can be used to define a temperature scale:

1. Pressure readings for **ice point** and **boiling point** of water are recorded.

2. Divide these readings into 100 divisions and call the ice point 0 and the
   boiling point 100 degrees. This is the **Celsius** scale.

3. But, it probably makes more sense to let zero on the temperature scale
   be that at zero pressure. So, we can extrapolate back to zero P and call
   this zero temperature. As shown later, (see page 27) this works if the
   volume is kept constant.

The ice point of water then turns
out to be 273.15 divisions from
zero and the boiling point 373.15
divisions from zero. This new
temperature scale is called
Kelvin and is the official SI
unit of temperature.

Thus, 0K = –273.15 °C
which is sometimes
called **absolute zero**.

0 K is 273.16
divisions below
triple point. Thus,
273.16 K = triple
point of water
which is 0.01 °C.

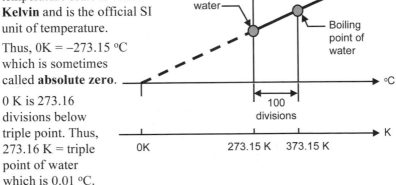

Pressure

Ice
point of
water

Boiling
point of
water

100
divisions

°C

0K   273.15 K   373.15 K   K

1.1.4 Examples

1. Give a short description of the following states of equilibrium:

   (a) Chemical equilibrium.
   (b) Mechanical equilibrium.
   (c) Thermal equilibrium.

Solution:

(a) Chemical equilibrium

   No chemical reactions are taking place inside the system boundary.

(b) Mechanical equilibrium

   Any external forces on the system boundary are in balance. The system is not accelerating or decelerating.

(c) Thermal equilibrium

   No heat flow is taking place within the system or across the system boundary.

2. Can a system be in a state of thermal equilibrium if the temperature of the system is different to that of the surroundings?

Solution:

Yes, but only if the system boundary is perfectly insulating. The condition for thermal equilibrium is that there is no *heat* flow within the system and across the system boundary. For the most part, this means no temperature gradients anywhere within the system and surroundings. If the system boundary is perfectly insulating, then the system and the surroundings can be at different temperatures and the system still be in a state of thermal equilibrium even though the system and the surroundings are not in thermal equilibrium with each other.

1.2 Heat and solids

Summary

- Energy exists in many forms.
- There is a flow, or transfer, of energy when a change of form takes place.
- **Heat** and **work** are words which refer to the amount of energy in transit from one place to another.
- Heat and work cannot be stored. Heat and work refer to **energy in transit**.

$\Delta L = \alpha L_o \Delta T$ Thermal expansion

$\Delta A = 2\alpha A_o \Delta T$

$\Delta V = \beta V_o \Delta T$

$\Delta Q = mc(T_2 - T_1)$ Specific heat

$C = M_m c$

$\Delta Q = nC(T_2 - T_1)$

$Q = mL_f$ Latent heat

$Q = mL_v$

1.2.1 Thermal expansion

When a solid is heated, there is an increase in
the linear dimensions as the temperature rises.

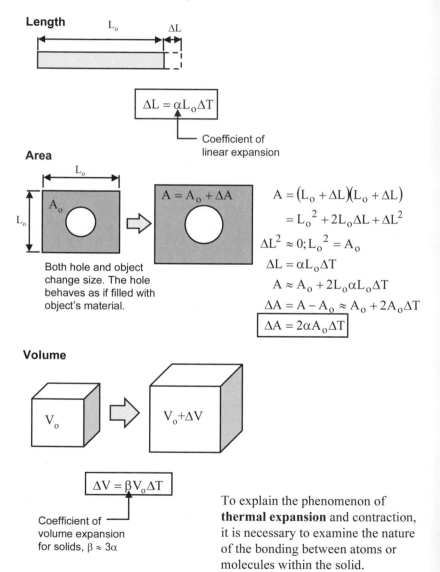

Length

$$\Delta L = \alpha L_o \Delta T$$

Coefficient of linear expansion

Area

$$A = A_0 + \Delta A$$

Both hole and object change size. The hole behaves as if filled with object's material.

$$A = (L_0 + \Delta L)(L_0 + \Delta L)$$
$$= L_0{}^2 + 2L_0\Delta L + \Delta L^2$$
$$\Delta L^2 \approx 0; L_0{}^2 = A_0$$
$$\Delta L = \alpha L_0 \Delta T$$
$$A \approx A_0 + 2L_0 \alpha L_0 \Delta T$$
$$\Delta A = A - A_0 \approx A_0 + 2A_0 \Delta T$$
$$\boxed{\Delta A = 2\alpha A_0 \Delta T}$$

Volume

$$V_o$$

$$V_o + \Delta V$$

$$\boxed{\Delta V = \beta V_o \Delta T}$$

Coefficient of volume expansion for solids, $\beta \approx 3\alpha$

To explain the phenomenon of **thermal expansion** and contraction, it is necessary to examine the nature of the bonding between atoms or molecules within the solid.

1.2.2 Atomic bonding in solids

Consider the forces acting between two atoms:

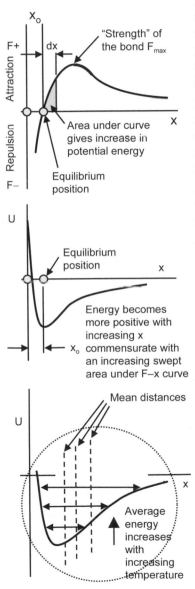

Note that *near the **equilibrium position***, the force required to move one atom away from another is very nearly directly proportional to the displacement x from the equilibrium position: **Hooke's law** or **linear elasticity**.

If one atom is held fixed, and the other is moved away by the application of an opposing force, then the area under the force-distance curve gives the increase in potential energy of the system.

As an atom is bought closer to another, the **potential energy** decreases until it reaches a minimum at the equilibrium position where the long range attraction is balanced by the short-range repulsion. Since **zero potential energy** is assigned at infinity, then the minimum at the equilibrium position corresponds to a negative potential energy.

The shape of the potential energy variation is not symmetric about the **minimum potential energy**.

The temperature, or internal energy is reflected by the amplitude of the oscillations and hence by the width of the trough in the energy distribution. Raising the temperature makes the average energy more positive. As well, due to the asymmetric nature of this distribution, the average or mean distance increases with increasing temperature leading to thermal expansion

1.2.3 Specific heat

1. When heat is transferred into a system, the temperature of the system (usually) rises.

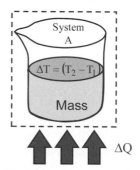

System
A

$\Delta T = (T_2 - T_1)$

Mass

ΔQ

Heat is energy in transit from one system to another

System B

2. The amount of energy (ΔQ) required to change the temperature of a mass of material is found to be dependent on:

• the mass of the body (m)   kg
• the temperature increase (ΔT) $^\circ$C, K
• the nature of the material (c)   J kg^{-1} K^{-1}

↓

Specific heat or
heat capacity

The **specific heat** is the amount of heat required to change the temperature of 1 kg of material by 1 $^\circ$C

Material	c (kJ kg^{-1} K^{-1})
Water	4.186
Steel	0.45
Cast iron	0.54
Aluminium	0.92

3. Putting these three things together in a formula, we have: $\Delta Q = mc(T_2 - T_1)$

Molar specific heat (C) is the amount of heat required to raise the temperature of <u>one mole</u> of the substance by 1 $^\circ$C

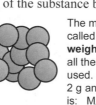

6.02 × 10^{23} molecules

$C = M_m c$

The mass (in kg) of one mole of substance is called the **molar mass** M_m. The **molecular weight** m.w. is the molar mass in grams. In all these formulas, molar mass M_m in kg is used. Hence, for water, where m.w. for H_2 = 2 g and m.w. for O_2 = 16 g, the molar mass is: M_m = (16+2)/1000
= 0.018 kg mol^{-1}.

Heat capacity formula: $\Delta Q = nC(T_2 - T_1)$

Number of moles ⎯⎯⎤ ⎣⎯⎯ Molar specific heat
(or **molar heat capacity**)

Note: We use small c for specific heat, large C for molar specific heat.

1.2.4 Latent heat

Latent heat is that associated with a **phase change**
e.g. Consider the heating of ice to water and then to steam:

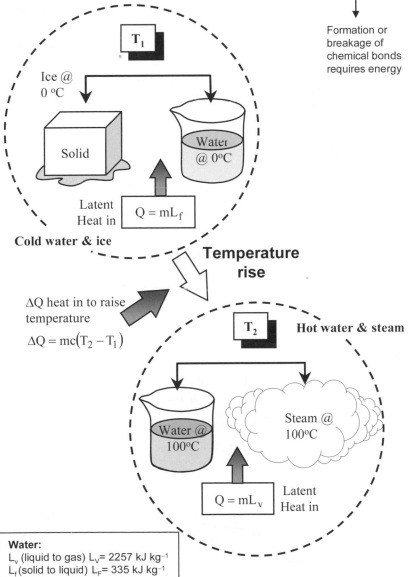

Formation or
breakage of
chemical bonds
requires energy

T_1

Ice @
0 °C

Solid

Water
@ 0°C

Latent
Heat in $Q = mL_f$

Cold water & ice

Temperature rise

ΔQ heat in to raise
temperature

$$\Delta Q = mc(T_2 - T_1)$$

T_2 **Hot water & steam**

Water @
100°C

Steam @
100°C

$Q = mL_v$ Latent
Heat in

Water:
L_v (liquid to gas) L_v= 2257 kJ kg^{-1}
L_f (solid to liquid) L_F= 335 kJ kg^{-1}

1.2.5 Examples

1. An aluminium pot of mass 0.6 kg contains 1.5 litres of water. Both
    are initially at 15 °C.
    (a) Calculate how many joules are required to raise both the saucepan
        and the water to 100 °C.
    (b) Calculate how many joules are required to change the water from
        liquid at 100 °C to steam at 100 °C.
    (c) What is the total energy needed to boil all the water away ?

Solution:

(a) $\Delta Q_{W_{15-100}} = 1.5(4.186)(100-15)$ Water from 15 °C to 100 °C

$= 533.7\text{kJ}$ Note: 1.5 L water = 1.5 kg

$\Delta Q_{Al_{15-100}} = 0.6(0.920)(100-15)$ Aluminium from 15 °C to 100 °C

$= 46.9\text{kJ}$

$Q_{Al+w} = 533.7 + 46.9$

$= 580.6\text{kJ}$

(b) $\Delta Q_{w-v} = 1.5(2.257)$

$= 3.385\text{MJ}$ Boil water (liquid to vapour)

(c) $Q_{Total} = 533.7 \times 10^3 + 3.385 \times 10^6 + 46.9 \times 10^3$

$= 3.97\text{MJ}$ Total energy required.

2. 0.5 kg of ice at –5 °C is dropped into 3 kg of water at 20 °C. Calculate
    the equilibrium temperature (neglecting external heat exchanges).

Solution:

With these types of problems, it is easiest to
equate all the terms involving a transfer of heat
to zero. In this problem, there are no external
heat exchanges, thus, the heat lost by the water
is equal to the heat gained by the ice.

Data:
c_{water} = 4186 J kg^{-1} K^{-1}
c_{Al} = 920 J kg^{-1} K^{-1}
L_v = 22.57 × 10^5 J kg^{-1}
L_f = 3.34 × 10^5 J kg^{-1}
ρ_{water} = 1000 kg m^{-3}
c_{ice} = 2110 J kg^{-1} K^{-1}

$Q_{water} = 3(4186)(T_2 - 20)$

$Q_{ice} = 0.5(2110)(0+5) + 0.5(3.34 \times 10^5) + 0.5(4186)(T_2 - 0)$

$0 = Q_{ice} + Q_{water}$

$0 = 5275 + 167000 + 2095T_2 + 12570T_2 - 251400$

$251400 = 172275 + 14665T_2$

$T_2 = 5.4°C$

1.3 Heat transfer

Summary

$$\dot{Q} = \frac{kA}{L}\left(T_{hot} - T_{cold}\right)$$ Heat flow by conduction

$$\frac{dQ}{dt} = -kA\frac{dT}{dx}$$ General heat conduction

$$R_{total} = \left(\frac{1}{C_a} + \frac{1}{C_b} + \frac{1}{C_c}\right)$$ Conduction - composite wall

$$\dot{Q} = \frac{T_2 - T_1}{R_{Total}}$$

$$\dot{Q} = \frac{-2\pi kL\left(T_2 - T_1\right)}{\ln\frac{r_2}{r_1}}$$ Conduction - radial pipe

$$\dot{Q} = hA(T_2 - T_1)$$ Convection

$$\frac{dT}{dt} = -\frac{kA}{mcL}(T_o - T_S)$$ Rate of cooling

$$\dot{Q}_e = e\sigma AT^4$$ Stefan–Boltzmann law

$$\Delta\dot{Q} = e_2\sigma A\left(T_2^4 - T_1^4\right)$$ Radiative heat transfer

$$T = \left(\frac{\dot{Q}_i}{A}\frac{1}{\sigma}\right)^{1/4}$$ Equilibrium temperature

1.3.1 Conduction

1. **Conduction** is microscopic transfer of heat. More energetic molecules transfer internal energy to less energetic molecules through collisions. Energy is transferred, not the molecules, from one point to another.

2. The rate of heat flow (J s^{-1}) is found to be dependent on:

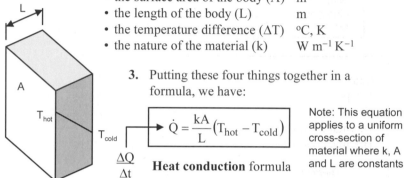

	Units
• the surface area of the body (A)	m^2
• the length of the body (L)	m
• the temperature difference (ΔT)	°C, K
• the nature of the material (k)	W m^{-1} K^{-1}

3. Putting these four things together in a formula, we have:

$$\dot{Q} = \frac{kA}{L}(T_{hot} - T_{cold})$$

$\frac{\Delta Q}{\Delta t}$ **Heat conduction** formula

Note: This equation applies to a uniform cross-section of material where k, A and L are constants

For the formulas to work, we must keep the signs straight and be aware of what *direction* we mean as being positive.

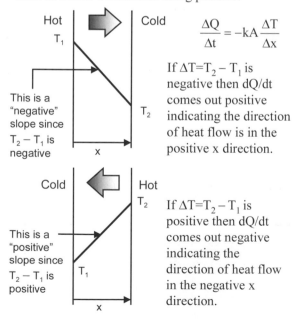

Hot
T_1 Cold

This is a "negative" slope since $T_2 - T_1$ is negative   x T_2

$$\frac{\Delta Q}{\Delta t} = -kA\frac{\Delta T}{\Delta x}$$

If $\Delta T = T_2 - T_1$ is negative then dQ/dt comes out positive indicating the direction of heat flow is in the positive x direction.

If the differences Δ are made very small, then we have:

$$\frac{dQ}{dt} = -kA\frac{dT}{dx}$$

Note: This equation is more general and may apply to non-uniform cross-sections (e.g. where A = f(x)).

Cold   Hot
T_2

This is a "positive" slope since $T_2 - T_1$ is positive T_1 x

If $\Delta T = T_2 - T_1$ is positive then dQ/dt comes out negative indicating the direction of heat flow in the negative x direction.

Note: the thickness L of the body is now expressed in terms of an interval "Δx" on the x axis.

1.3.2 Thermal conductivity

Thermal conductivity (k)

Material	k (W m⁻¹ K⁻¹)
Aluminium	220
Steel	54
Glass	0.79
Water	0.65
Fibreglass	0.037
Air	0.034

- a material property
- units: $W\ m^{-1}\ K^{-1}$
- high value indicates good thermal **conductor**
- low value indicates good thermal **insulator**

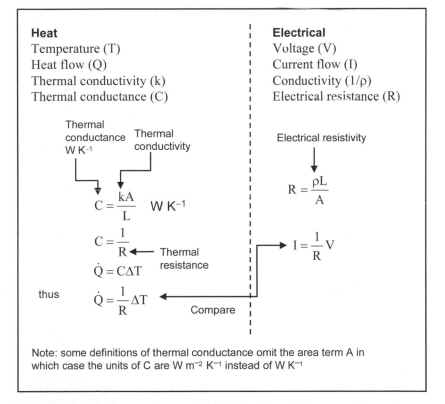

Heat
Temperature (T)
Heat flow (Q)
Thermal conductivity (k)
Thermal conductance (C)

Electrical
Voltage (V)
Current flow (I)
Conductivity (1/ρ)
Electrical resistance (R)

Thermal conductance W K⁻¹   Thermal conductivity

$$C = \frac{kA}{L} \quad W\ K^{-1}$$

$$C = \frac{1}{R} \longleftarrow \text{Thermal resistance}$$

$$\dot{Q} = C\Delta T$$

thus

$$\dot{Q} = \frac{1}{R}\Delta T$$

Electrical resistivity

$$R = \frac{\rho L}{A}$$

$$I = \frac{1}{R}V$$

Compare

Note: some definitions of thermal conductance omit the area term A in which case the units of C are W m⁻² K⁻¹ instead of W K⁻¹

Mathematically, temperature gradient is very much like potential gradient in electricity. Heat flow is similar to current flow.

1.3.3 Composite wall

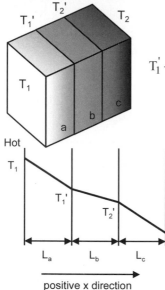

Hot

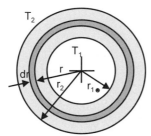

positive x direction

For slab (a)   (b)   (c)

$$\dot{Q} = C_a\left(T_1' - T_1\right)$$

$$T_1' - T_1 = \frac{\dot{Q}}{C_a} \qquad T_2' - T_1' = \frac{\dot{Q}}{C_b} \qquad T_2 - T_2' = \frac{\dot{Q}}{C_c}$$

Add (a)+(b)+(c) together:

$$T_2 - T_1 = \dot{Q}\left(\frac{1}{C_a} + \frac{1}{C_b} + \frac{1}{C_c}\right)$$

$$\frac{1}{C_{Total}} = R_{Total} = R_a + R_b + R_c$$

Thermal resistances in series add just like electrical resistors in series.

$$\dot{Q} = \frac{T_2 - T_1}{R_{Total}}$$

Composite wall formula

The **radial pipe** problem is different to the case of a block because the heat flows from the inside of the pipe to the outside across an ever expanding cross-sectional area A. In the case of the block, the area A through which the heat flowed was a constant (did not change with x). Here the area A is a function of r.

$$\dot{Q} = -kA\frac{dT}{dx}$$

$$A = 2\pi rL$$

$$\dot{Q} = -k(2\pi rL)\frac{dT}{dr}$$

Note: If T_2-T_1 is positive then dQ/dt is negative (i.e. opposite to direction of positive r)

For heat flow Q in terms of T_1, T_2 and r_1 and r_2, we must integrate the effect of the increasing value of A with respect to r.

$$\frac{1}{\dot{Q}}dT = \frac{-1}{2k\pi rL}dr$$

$$\frac{1}{\dot{Q}}\int_{T_1}^{T_2} dT = \frac{-1}{2k\pi L}\int_{r_1}^{r_2}\frac{1}{r}dr$$

$$\frac{(T_2 - T_1)}{\dot{Q}} = \frac{-1}{2k\pi L}\ln\frac{r_2}{r_1}$$

Radial conduction formula

$$\dot{Q} = \frac{dQ}{dt} = \frac{-2\pi kL(T_2 - T_1)}{\ln\frac{r_2}{r_1}}$$

1.3.4 Rate of cooling

Question: If heat flows from a body (by conduction) and the body cools, what is the rate at which the temperature of the body changes (dT/dt)?

Answer: Start with heat capacity formula: $\Delta Q = mc\Delta T$

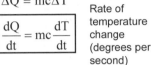

Differentiate w.r.t. time to get (assuming no change of state)

$$\frac{dQ}{dt} = mc\frac{dT}{dt}$$

Rate of temperature change (degrees per second)

But, for conduction, the rate of extraction of heat is:

Negative indicates a temperature *decrease* w.r.t. time

$$\frac{dQ}{dt} = -\frac{kA}{L}(T - T_S)$$

→ Surroundings

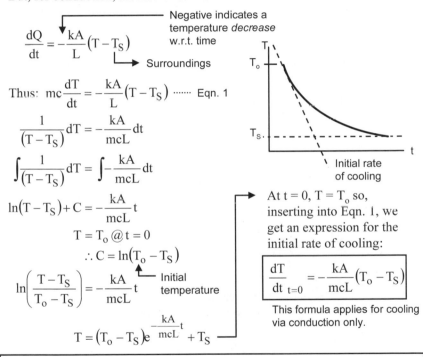

Thus: $mc\dfrac{dT}{dt} = -\dfrac{kA}{L}(T - T_S)$ ⋯⋯ Eqn. 1

$$\frac{1}{(T - T_S)}dT = -\frac{kA}{mcL}dt$$

$$\int\frac{1}{(T - T_S)}dT = \int -\frac{kA}{mcL}dt$$

$$\ln(T - T_S) + C = -\frac{kA}{mcL}t$$

$$T = T_0 @ t = 0$$

$$\therefore C = \ln(T_0 - T_S)$$

$$\ln\left(\frac{T - T_S}{T_0 - T_S}\right) = -\frac{kA}{mcL}t$$

Initial temperature

$$T = (T_0 - T_S)e^{-\frac{kA}{mcL}t} + T_S$$

At t = 0, T = T$_0$ so, inserting into Eqn. 1, we get an expression for the initial rate of cooling:

$$\boxed{\left.\frac{dT}{dt}\right|_{t=0} = -\frac{kA}{mcL}(T_0 - T_S)}$$

This formula applies for cooling via conduction only.

A similar analysis can be done for a generalised heat transfer coefficient K that takes into account conduction, convection and radiation. This empirical relationship is known as **Newton's law of cooling**.

$$\left.\frac{dT}{dt}\right|_{t=0} = -K(T - T_S)$$

$$\frac{dT}{dt} = -K(T_0 - T_S)$$

If the initial rate of cooling is measured, then the heat transfer coefficient K can be calculated: —

$$K = -\frac{dT}{dt}\frac{1}{(T_0 - T_S)}$$

1.3.5 Convection

Fluid (in this case, air) loses heat as it cools (decrease in temperature)

Convection current

Fluid travels from one place to another:

- Due to density change - Natural or free **convection**
- Due to mechanical assistance (pump or fan) - **forced convection**

Fluid (e.g. air) absorbs heat (increase in temperature)

The fluid takes heat with it, thus the flow of the fluid represents a heat flow.

The effectiveness or measure of heat flow by convection is given by the **convective heat transfer coefficient**. This is not entirely a material property, but depends also upon the circumstances of the heat flow.

$$\dot{Q} = hA(T_2 - T_1)$$

└─ **Convective heat transfer coefficient** depends on:

- Fluid properties: thermal conductivity, density, viscosity, specific heat, expansion coefficient

Prandtl number ──▶ $Pr = \dfrac{\eta c_p}{k}$

- Viscosity (kg m⁻¹ s⁻¹) → Viscosity $(kg\ m^{-1}\ s^{-1})$
- Specific heat
- Thermal conductivity

- Flow characteristics: whether natural or forced, surface geometry, flow regime (laminar or turbulent)

Reynolds number ──▶ $Re = \dfrac{vL\rho}{\eta}$

- Velocity of fluid
- Density
- Viscosity

The **Prandtl number** and **Reynolds number** are dimensionless quantities that combine to form the convective heat transfer coefficient h. The way in which they are combined depends on the circumstances of the flow. The final value of convective heat transfer coefficient is usually obtained by experiment under controlled conditions. From the equations above, we can see that convection is assisted with a large velocity (v), and high density, high specific heat and low thermal conductivity.

1.3.6 Thermal radiation

The transfer of energy by **electromagnetic waves** in the thermal range. Let us call energy transferred by electromagnetic radiation: **radiant energy.**

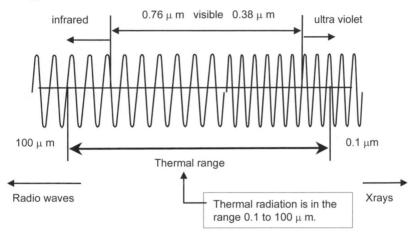

When radiant energy coming from a source falls on body, part of it is absorbed, part of it may be reflected, and part transmitted through it.

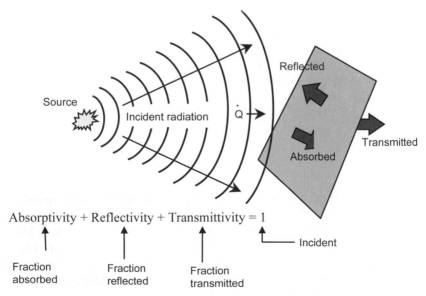

Absorptivity + Reflectivity + Transmittivity = 1

1.3.7 Emission

Any body whose temperature is above absolute zero emits radiant energy according to the Stefan–Boltzmann law.

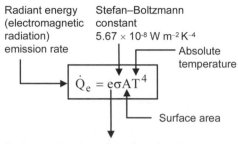

Radiant energy
(electromagnetic
radiation)
emission rate

Stefan–Boltzmann
constant
5.67×10^{-8} W m^{-2} K^{-4}

Absolute temperature

$$\dot{Q}_e = e\sigma A T^4$$

Surface area

e is the **emissivity** of surface (varies between 0 and 1) and depends on:

- nature of surface
- temperature of surface
- wavelength of the radiation being emitted or absorbed

Material	e
Polished aluminium	0.095
Oxidised aluminium	0.20
Water	0.96
Black Body	1.0

e indicates how well a body *emits or absorbs* radiant energy. A good emitter is a good absorber. Emissivity is sometimes called **relative emittance**.

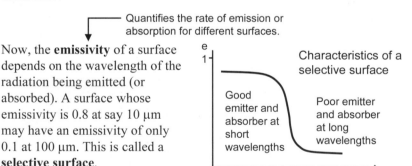

Quantifies the rate of emission or absorption for different surfaces.

Now, the **emissivity** of a surface depends on the wavelength of the radiation being emitted (or absorbed). A surface whose emissivity is 0.8 at say 10 µm may have an emissivity of only 0.1 at 100 µm. This is called a **selective surface**.

Characteristics of a selective surface

Good emitter and absorber at short wavelengths

Poor emitter and absorber at long wavelengths

Despite these variations in emissivity, for any particular wavelength, it so happens that the **absorptivity** α of a surface is equal to its **emissivity** e for a surface in equilibrium. This is known as **Kirchhoff's radiation law**.

1.3.8 Absorption and emission

We might well ask the question "What happens to the energy that is absorbed?"

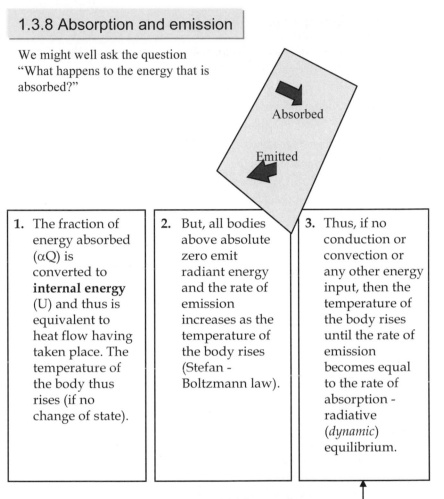

Absorbed

Emitted

1. The fraction of energy absorbed (αQ) is converted to **internal energy** (U) and thus is equivalent to heat flow having taken place. The temperature of the body thus rises (if no change of state).

2. But, all bodies above absolute zero emit radiant energy and the rate of emission increases as the temperature of the body rises (Stefan - Boltzmann law).

3. Thus, if no conduction or convection or any other energy input, then the temperature of the body rises until the rate of emission becomes equal to the rate of absorption - radiative (*dynamic*) equilibrium.

If this didn't happen, then the body would continue to absorb radiation and the temperature would rise indefinitely till melt-down!

1.3.9 Radiative heat transfer

1. Suppose the surroundings (1) are maintained at a low temperature T_1 and the body (2) maintained at a higher temperature T_2. e.g. by electricity.

2. All bodies simultaneously emit and absorb radiant energy.

3. The rate of emission from the body is given by: $\dot{Q}_e = e_2 \sigma A T_2^4$

4. The rate of absorption by the body is given by $\dot{Q}_a = \alpha \dot{Q}_i$ and depends on:

   • the temperature of the **surroundings**
   • the emissivity (or absorptivity) of the body

$\dot{Q}_a = \alpha \dot{Q}_i = \alpha e_1 \sigma A T_1^4 = e_2 e_1 \sigma A T_1^4$ where $\dot{Q}_i = e_1 \sigma A T_1^4$ since $\alpha = e$

Consider the energy flow into and out from the body:

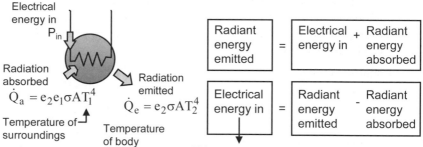

Electrical energy in P_{in}

Radiation absorbed $\dot{Q}_a = e_2 e_1 \sigma A T_1^4$

Temperature of surroundings

Radiation emitted $\dot{Q}_e = e_2 \sigma A T_2^4$

Temperature of body

Radiant energy emitted	=	Electrical energy in	+	Radiant energy absorbed

Electrical energy in	=	Radiant energy emitted	−	Radiant energy absorbed

At radiative dynamic equilibrium, **ENERGY OUT = ENERGY IN**

This quantity represents the *net* heat flow ΔQ from the body and is equal to the rate of electrical energy that must be supplied to maintain the body at T_2.

$$\Delta\dot{Q} = e_2 \sigma A T_2^4 - e_2 e_1 \sigma A T_1^4$$

If the surroundings have an emissivity of 1, then $e_1 = 1$ then we have :

$$\boxed{\Delta\dot{Q} = e_2 \sigma A \left(T_2^4 - T_1^4 \right)}$$

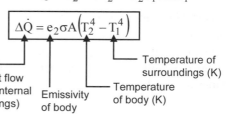

Net Heat flow (positive in this case indicates net radiant heat flow *from* the body and transfer of internal energy from body to surroundings)

Emissivity of body

Temperature of body (K)

Temperature of surroundings (K)

1.3.10 Radiation emission

Radiant energy is emitted or absorbed in the **thermal range** - 0.1 to 100 μm.

As temperature increases:
- total energy (area under curve) increases
- peak in emission spectrum shifts to shorter wavelengths
- can only be explained using quantum theory

Within this range, the amount of energy emitted is not uniformly distributed.

For a **black-body**, e =1 no matter what the wavelength. The emission spectrum for a black body is the maximum possible for any given temperature and has the shape which can be

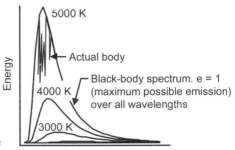

5000 K

Actual body

Black-body spectrum. e = 1 (maximum possible emission) over all wavelengths

4000 K

3000 K

Energy

Wavelength

calculated using quantum theory. For an actual body, e < 1 and varies depending on the wavelength being emitted. This leads to deviations from the black-body emission curve.

Variation of emission with distance

If the source of radiation may be regarded as a **point source** (due to the source being small or distance to object being large) then the energy reaching a perpendicular surface varies inversely as the square of the distance.

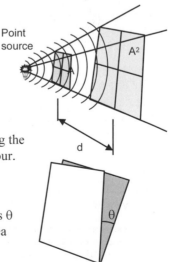

Point source

A²

d

If distance d is doubled, then energy reaching the surface in W m⁻² is reduced by a factor of four.

Variation of emission with angle

If the surface is not perpendicular, then the radiation intensity (W m⁻²) is reduced by cos θ – i.e. the radiation is spread over a larger area and thus the intensity is reduced.

θ

1.3.11 Equilibrium temperature

Consider an insulated plate with incident
radiation. If the temperature of the plate is
neither increasing or decreasing, then, in
the absence of any other energy input, rate
of absorption = rate of emission.

$$\dot{Q}_a = \alpha \dot{Q}_i$$

Radiative equilibrium

At equilibrium: $\dot{Q}_a = \dot{Q}_e$

$$\dot{Q}_e = e\sigma AT^4$$

$$\alpha \dot{Q}_i = e\sigma AT^4$$

assuming $e_{e\lambda} = \alpha_{a\lambda}$

at this condition

then $\dot{Q}_i = \sigma AT^4$

$$T = \left(\frac{\dot{Q}_i}{A} \frac{1}{\sigma} \right)^{1/4}$$

The final **steady-state** temperature depends on radiation intensity (W m^{-2})
falling on the body and not on the emissivity of the surface! i.e., a black
surface and a white surface reach the same equilibrium temperature.

Now, $\alpha = e$ is true *for any particular
wavelength* of radiation we are talking
about. But, the sun, being very hot,
transmits a lot of energy at short
wavelengths. Objects placed in the sun
receive a portion of this energy with a
certain value of α.

This does not necessarily mean
temperature of plate is equal to
temperature of source. If the plate
receives *all* the energy emitted
from the source and the source is
a black-body, then temperatures
are equal at radiative equilibrium.

As an object heats up, it begins to radiate more and more energy and the
rate of emission depends upon T^4. Thus, the temperature reaches a point
where the rate of emission = rate of absorption. BUT!, the rate of emission
also depends on the value of e as well as T^4. Generally, e and α both vary
with wavelength, so if a body has a high value of α for short wavelengths,
and a low value of e at long wavelengths (e.g. a black-painted car) then the
body would have to reach a higher steady state temperature to keep the rate
of absorption = rate of emission compared to the case where $\alpha = e$.

People with dark skin absorb and emit radiation at a greater rate than people
with light skin. But for skin, unlike paint, the absorptivity at short wavelengths is
less than the emissivity at long wavelengths. Hence the balance is for a lower
equilibrium temperature compared to the case where $\alpha = e$ for both short and
long wavelengths.

1.3.12 Examples

1. Calculate the heat flow through a glass sliding door of a house if it is 25 °C inside and 5 °C outside and the door has an area of 1.9 m^2 and thickness 3 mm where k_{glass} = 0.79 W m^{-1} K^{-1}.

Solution:

$$\dot{Q} = \frac{kA(T_2 - T_1)}{L}$$
$$= \frac{0.79(1.9)(25-5)}{0.003}$$
$$= 10.0\,kW$$

2. Calculate the temperature of a steel block 60 seconds after being heated to 500 °C and then allowed to cool by being placed on a steel table top maintained at 20 °C. The contact surface 20 × 20 mm, the thickness is 20 mm, the specific heat is 0.45 kJ kg^{-1}K^{-1}, the mass is 0.125 kg and the coefficient of thermal conductivity is 54 W m^{-1} K^{-1}. Ignore any cooling by convection or radiation.

Solution:

$$T = (T_o - T_S)e^{-\frac{kA}{mcL}t} + T_S$$
$$= (500 - 20)\exp\left[-\frac{54(0.0004)}{0.125(450)0.02}60\right] + 20$$
$$= 173.6°C$$

3. The oil sump on a motor vehicle is made from an alloy casting with fins to assist in cooling. If the fins have a total surface area of 600 cm^2 in addition to the flat surface area of the casting also of 600 cm^2, and the oil temperature is 85 °C, determine the rate of heat dissipation from the sump casting when the vehicle is stopped and h = 10 W m^{-2} K^{-1} and the ambient air temperature is 30 °C.

Solution:

$$\dot{Q} = hA(T_2 - T_1)$$
$$= 10(1200 \times 10^{-4})(85 - 30)$$
$$= 66\,W$$

4. A bright chrome seat belt buckle rests on the seat of a parked car and receives radiation of intensity 0.75 kW m^{-2} from the midday sun.

    (a) Calculate the rate of energy absorption per m^2 of surface area of the buckle per second (emissivity of chrome: 0.07).

    (b) Determine an expression for the rate of energy emission per m^2 of surface area and hence calculate the equilibrium temperature of the buckle (neglect conduction and convection and any variations of emissivity with wavelength).

    (c) Calculate the equilibrium temperature of the buckle if it were painted with black paint (emissivity of black painted steel: 0.85) - justify your answer.

      Stefan-Boltzmann constant: $\sigma = 5.67 \times 10^{-8}$ W m^{-2} K^{-4}

Solution:

$$I = 0.75\text{kWm}^{-2}$$

$$\varepsilon = 0.07$$

$$= \alpha$$

Consider 1 m^2 of area:

$$\dot{Q}_a = \alpha(I)$$

$$= 0.07(750)$$

$$= 52.5\text{W}$$

$$\dot{Q}_e = \varepsilon\sigma A T^4$$

$$\dot{Q}_e = 0.07\left(5.67 \times 10^{-8}\right)T^4$$

$$\dot{Q}_e = \dot{Q}_a$$

$$T^4 = \frac{0.07(750)}{0.07\left(5.67 \times 10^{-8}\right)}$$

$$= \frac{52.5}{3.97 \times 10^{-9}}$$

$$T = 339\text{K}$$

$$= 66°\text{C}$$

Equilibrium temperature would be the same.
Emissivity e = 0.07 cancels out in expression for T

1.4 Gases

Summary

$$p = p_g + p_{atm}$$ Absolute pressure

$$p = p_A + p_B$$ Partial pressures

$$\frac{p_1 V_1}{T_1} = \frac{p_2 V_2}{T_2}$$ Combined gas law

$$p_1 V_1 = p_2 V_2$$ Boyle's law

$$\frac{V_1}{T_1} = \frac{V_2}{T_2}$$ Charles' law

$$pV = nRT$$ Equation of state

1.4.1 Solids, liquids and gases

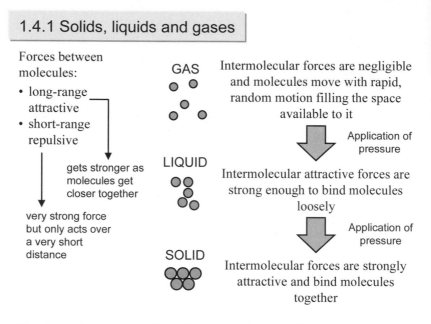

Forces between molecules:
- long-range attractive
- short-range repulsive

gets stronger as molecules get closer together

very strong force but only acts over a very short distance

GAS

Intermolecular forces are negligible and molecules move with rapid, random motion filling the space available to it

Application of pressure

LIQUID

Intermolecular attractive forces are strong enough to bind molecules loosely

Application of pressure

SOLID

Intermolecular forces are strongly attractive and bind molecules together

The thermal properties of a perfect or **ideal gas** are the most convenient to study. The properties of an ideal gas are:

- The molecules of the gas occupy a very small volume compared to the volume of the container
- The molecules are very distant from one another relative to their size and only interact during collisions
- Collisions between molecules are elastic

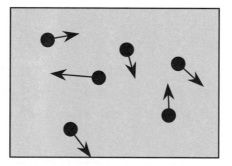

Real gases often behave like ideal gases at conditions in which no phase changes occur.

1.4.2 Pressure

One of the most important macroscopic properties of a gas is its **pressure**.

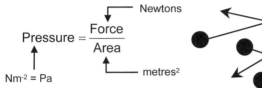

$$\text{Pressure} = \frac{\text{Force}}{\text{Area}}$$

Newtons

$Nm^{-2} = Pa$

metres^2

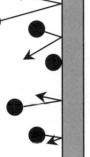

Impact of a molecule on the wall of the container exerts a force on the wall. There are many such impacts per second. The total force per unit area is called **pressure**. Pressure is the average effect of the many impacts resulting from molecule to wall collisions.

It is important to realise that pressure gauges measure the pressure above or below atmospheric pressure. Thus in thermodynamic formulae, absolute pressure must always be used.

$$p = p_g + p_{atm}$$

Absolute zero of pressure is no pressure at all. Pressures above zero pressure are called absolute pressures.

A pressure gauge usually measures the pressure above or below atmospheric pressure

1 atmosphere:
= 760 Torr
= 1013 millibars
= 101.3 kPa
= 760 mm Hg

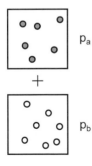

p_a

p_b

In a mixture of gases, the total pressure is the sum of the pressures of the component gases if those gases occupied the volume of the container on its own. These pressures are called **partial pressures**.

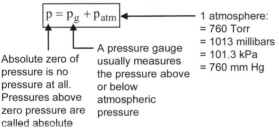

Dry air	%Vol	Partial pressure (mbar)
N_2	78.08	791.1
O_2	20.95	212.2
Ar	0.93	9.4
CO_2	0.03	0.3
		Total: 1013 mbar

$+$

$=$

p_{total}

1.4.3 Gas laws

Macroscopic properties of a gas

- Pressure
- Temperature
- Volume
- Mass

These quantities specify the **state** of a gas

Consider a mass (m) of gas:

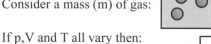

If p, V and T all vary then:

$$\frac{p_1 V_1}{T_1} = \frac{p_2 V_2}{T_2}$$ **Combined gas law**

Can be obtained from kinetic theory or from experiment.

If temperature T is a constant:

$$p_1 V_1 = p_2 V_2$$ **Boyle's law**

If pressure p is a constant:

$$\frac{V_1}{T_1} = \frac{V_2}{T_2}$$ **Charles' law**

Notes: These laws cannot be applied when the mass of gas changes during the process. Pressures and temperatures are absolute.

Let us express the mass of a gas indirectly by specifying the number of moles:

No. moles → $n = \dfrac{m}{M_m}$

6.02 × 10²³ molecules

Mass in kg

Molar mass

Experiment shows that Boyle's law and Charles law leads to:

By using moles, we get the ideal gas equation with the universal gas constant R (units J mol⁻¹K⁻¹). otherwise, value for R depends on the nature of the gas (i.e. no longer universal) and has units J kg⁻¹K⁻¹ .

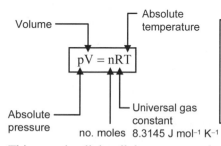

Volume

Absolute temperature

$$pV = nRT$$

Absolute pressure

Universal gas constant

no. moles 8.3145 J mol⁻¹ K⁻¹

Example: Calculate the volume occupied by one mole of an ideal gas at 273 K at atmospheric pressure.

$$pV = nRT$$
$$101.3(V) = 1(8.314)(273)$$
$$V = 22.406L$$

This equation links all the macroscopic quantities needed to describe the (steady) state of an *ideal* gas and is thus called an **equation of state**. There are other equations of state, mostly used to describe the state of **real gases**.

1.4.4 Phases of matter

In a **p–V diagram**, the temperature is kept constant, volume decreased and pressure recorded.

Vapour pressure
The partial pressure exerted by the vapour when it is in equilibrium with its liquid. It depends on temperature and nature of the substance. The temperature at which the vapour pressure equals the prevailing atmospheric pressure is called the **boiling point**.

Critical Temperature T_c
There is, for each gas, a temperature above which the attractive forces between molecules are not strong enough to produce liquefaction no matter how high a pressure is applied.

T_c H_2O = 647 K at 218 atm
T_c He = 5.2 K at 2.3 atm

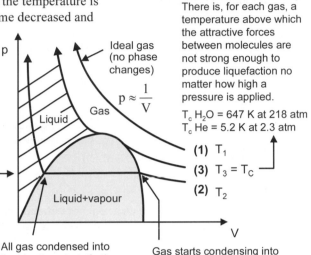

All gas condensed into liquid, attempts to further reduce volume produce large increase in pressure as liquid is compressed.

Gas starts condensing into liquid, no change in pressure as volume decreases

In a **phase diagram**, we keep the volume V a constant and plot pressure vs temperature.

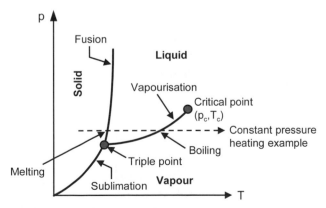

At each point (p,T) only a single phase can exist except on the lines where there is phase equilibrium. At the triple point, solid, liquid and vapour exist together in equilibrium.

1.4.5 Examples

1. A motorist checks the pressure of his tyres after driving at high speed and measures 300 kPa. He notices that the temperature of the tyre is 50 °C. What would be the pressure when the tyre is at room temperature (assume the volume of the tyre has remained constant)?

Solution:

$$\frac{p_1}{T_1} = \frac{p_2}{T_2}$$

$$\frac{p_1}{293} = \frac{300}{323}$$

$$p_1 = 272\,kPa$$

2. Calculate the mass of air in a room of volume 200 m³ at 20 °C and 101.3 kPa given that the molecular mass of air is 28.92 g mol⁻¹

Solution:

$$pV = nRT$$
$$101300(200) = n(8.314)(20 + 273)$$
$$n = 8312\,moles$$
$$m = 8312(.02892)$$
$$= 240.5\,kg$$

3. An air compressor has a tank of volume 0.2 m³. When it is filled from atmospheric pressure, the pressure gauge attached to the tank reads 500 kPa after the tank has returned to room temperature (20 °C). Calculate the mass of air in the tank.

Solution:

$$pV = nRT$$
$$(101.3 + 500)(1000)(0.2) = n(8.314)(20 + 273)$$
$$n = 49.4\,moles$$
$$m = 49.4(.02892)$$
$$= 1.43\,kg$$

1.5 Work and thermodynamics

Summary

$$Q = mc(T_2 - T_1)$$ Specific heat

$$Q = nC(T_2 - T_1)$$ Molar specific heat

$$Q - W = (U_2 - U_1)$$ First law of thermodynamics

$$R = C_p - C_v$$ Universal gas constant

$$W = \int_{V_1}^{V_2} p(V)\,dV$$ Work done on or by a gas

1.5.1 Gas

Experiments show that when a gas is heated at constant volume, the **specific heat** c_v is always less than that if the gas is heated at constant pressure c_p.

For a given temperature rise ΔT, there will always be a volume change ΔV with the constant pressure process meaning that the energy into the system has to both raise the temperature *and* do work, thus c_p is always greater than c_v.

Air	
c_p = 1.005 kJ kg^{-1} K^{-1}	
c_v = 0.718 kJ kg^{-1} K^{-1}	

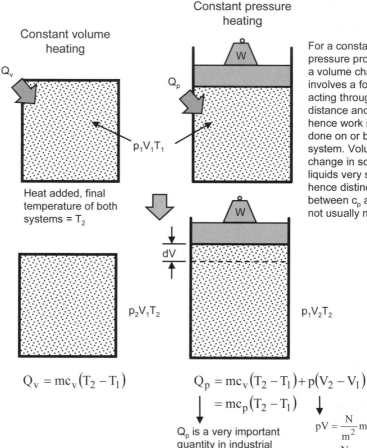

Constant volume heating

Q_v

$p_1V_1T_1$

Heat added, final temperature of both systems = T_2

$p_2V_1T_2$

Constant pressure heating

Q_p

W

For a constant pressure process, a volume change involves a force acting through a distance and hence work is done on or by the system. Volume change in solids & liquids very small hence distinction between c_p and c_v not usually made.

dV

W

$p_1V_2T_2$

$$Q_v = mc_v(T_2 - T_1)$$

$$Q_p = mc_v(T_2 - T_1) + p(V_2 - V_1)$$
$$= mc_p(T_2 - T_1)$$

Q_p is a very important quantity in industrial processes and is given the name **enthalpy**.

$$pV = \frac{N}{m^2}m^3$$
$$= N \cdot m$$
$$= J$$

1.5.2 1st law of thermodynamics

For the **constant pressure** process:

$$Q_p = mc_v(T_2 - T_1) + p(V_2 - V_1)$$

$$= mc_p(T_2 - T_1)$$

$$\boxed{mc_p(T_2 - T_1)} - \boxed{p(V_2 - V_1)} = \boxed{mc_v(T_2 - T_1)}$$

↑ Heat in or out of gas

↑ Work done on or by gas

↑ Change in internal energy

Internal energy is the kinetic energy of vibration of the molecules that make up the gas.

For ALL processes:

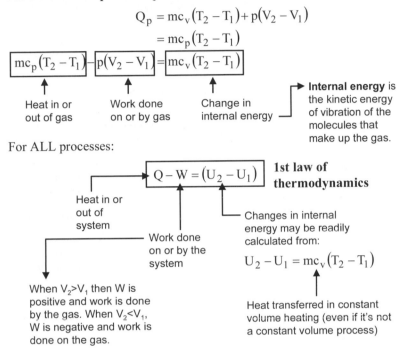

$$\boxed{Q - W = (U_2 - U_1)}$$ **1st law of thermodynamics**

Heat in or out of system

Work done on or by the system

Changes in internal energy may be readily calculated from:

$$U_2 - U_1 = mc_v(T_2 - T_1)$$

When $V_2 > V_1$ then W is positive and work is done by the gas. When $V_2 < V_1$, W is negative and work is done on the gas.

Heat transferred in constant volume heating (even if it's not a constant volume process)

Consider a constant pressure process $p_1 = p_2$ and the 1st law.

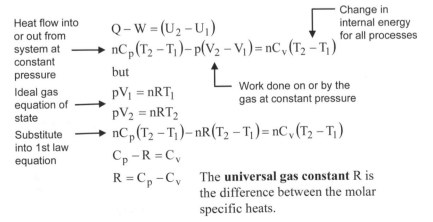

Heat flow into or out from system at constant pressure

Ideal gas equation of state

Substitute into 1st law equation

Change in internal energy for all processes

$$Q - W = (U_2 - U_1)$$

$$nC_p(T_2 - T_1) - p(V_2 - V_1) = nC_v(T_2 - T_1)$$

but

Work done on or by the gas at constant pressure

$$pV_1 = nRT_1$$

$$pV_2 = nRT_2$$

$$nC_p(T_2 - T_1) - nR(T_2 - T_1) = nC_v(T_2 - T_1)$$

$$C_p - R = C_v$$

$$R = C_p - C_v$$ The **universal gas constant** R is the difference between the molar specific heats.

1.5.3 p–V diagram

For a system at equilibrium, the properties are the same throughout the system. For a gas, only two independent properties are required (e.g. p and V). The equation of state gives the connection between these and the others (e.g. T and n).

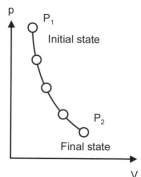

If the change of state from P_1 to P_2 occurs as a series of small steps, each of which represents an equilibrium condition, then a line joining the two points represents an equilibrium or **quasi-static** process. If the system undergoes a process, the state of the gas may change from the initial state to the final state. In a quasi-static process, the out of balance condition that drives the change of state is very small so that there are no "dynamic" effects influencing the process.

On a p–V diagram, the area under the curve between P_1 and P_2 is

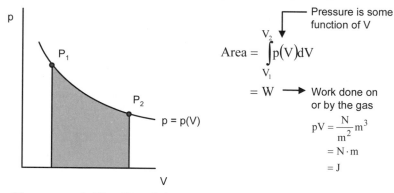

Pressure is some function of V

$$\text{Area} = \int_{V_1}^{V_2} p(V) dV$$

$$= W \longrightarrow \text{Work done on or by the gas}$$

$$pV = \frac{N}{m^2} m^3$$

$$= N \cdot m$$

$$= J$$

- If gas expands $V_2 - V_1 > 0$ then work is done by the gas
- If gas contracts from $V_2 - V_1 < 0$ then work is done on the gas

1.5.4 Example

1. A milkshake is prepared by mixing the 500 g of milk and ice cream in an electric blender. If the blender has a power rating of 100 W, and the initial temperature of the milk and ice cream were 2 °C, calculate the temperature of the mixture after being mixed for 3 minutes (ignore heat flow from the surroundings and assume the mixture has the specific heat the same as water).

Solution:

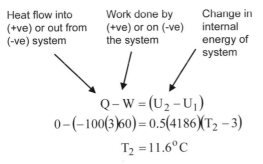

Heat flow into (+ve) or out from (-ve) system

Work done by (+ve) or on (-ve) the system

Change in internal energy of system

$$Q - W = (U_2 - U_1)$$
$$0 - (-100(3)60) = 0.5(4186)(T_2 - 3)$$
$$T_2 = 11.6 °C$$

Note, in this example, it is seen that the first law applies to liquids (and solids) as well as gases. However, here, work is done *on* the liquid, thus W is actually negative (which is consistent with V_2-V_1 < 0 when work is done on a gas).

1.6 Gas processes

Summary

$$\frac{V_1}{T_1} = \frac{V_2}{T_2}$$ Constant pressure

$$\Delta U = nC_v(T_2 - T_1)$$

$$W = p(V_2 - V_1)$$

$$Q = nC_p(T_2 - T_1)$$

$$\frac{p_1}{T_1} = \frac{p_2}{T_2}$$ Constant volume

$$\Delta U = nC_v(T_2 - T_1)$$

$$W = 0$$

$$Q = nC_V(T_2 - T_1)$$

$$p_1V_1 = p_2V_2$$ Constant temperature

$$\Delta U = 0$$

$$W = p_1V_1 \ln\frac{V_2}{V_1}$$

$$Q = W$$

$$p_1V_1^n = p_2V_2^n$$ Polytropic

$$\Delta U = nC_v(T_2 - T_1)$$

$$W = \frac{p_1V_1 - p_2V_2}{n-1}$$

$$Q = nC_n(T_2 - T_1)$$ Adiabatic

$$C_n = \frac{C_p - nC_v}{n-1} \qquad n = \gamma = \frac{C_p}{C_v}$$

1.6.1 Gas processes

If a thermodynamic system undergoes a process, the state of the gas may change from the initial state to the final state. The process may occur at a constant volume, temperature or pressure, or all of these may vary.

A gas process at **constant pressure** is called **isobaric**.

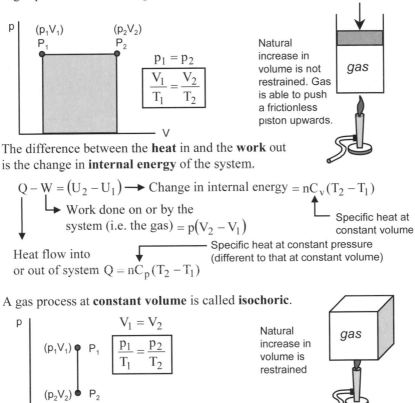

$$p_1 = p_2$$

$$\frac{V_1}{T_1} = \frac{V_2}{T_2}$$

Natural increase in volume is not restrained. Gas is able to push a frictionless piston upwards.

The difference between the **heat** in and the **work** out is the change in **internal energy** of the system.

$$Q - W = (U_2 - U_1) \longrightarrow \text{Change in internal energy} = nC_v(T_2 - T_1)$$

Work done on or by the system (i.e. the gas) $= p(V_2 - V_1)$

Specific heat at constant volume

Heat flow into or out of system $Q = nC_p(T_2 - T_1)$

Specific heat at constant pressure (different to that at constant volume)

A gas process at **constant volume** is called **isochoric**.

$$V_1 = V_2$$

$$\frac{p_1}{T_1} = \frac{p_2}{T_2}$$

Natural increase in volume is restrained

Heat flow is equal to change in **internal energy** and no **work** is done on or by the system.

$$Q - W = (U_2 - U_1) \longrightarrow \text{Change in internal energy} = nC_v(T_2 - T_1)$$

Work done on or by the system (i.e. the gas) $= \int_{V_1}^{V_2} p dV = 0$

Specific heat at constant volume

Heat flow into or out of system $Q = nC_V(T_2 - T_1)$

A gas process at **constant temperature** is called **isothermal**.

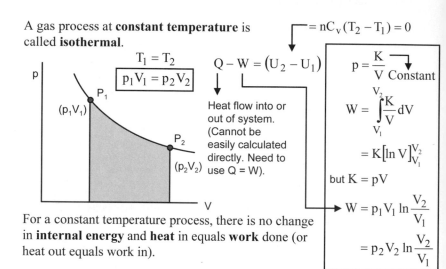

For a constant temperature process, there is no change in **internal energy** and **heat** in equals **work** done (or heat out equals work in).

A gas process in which p, V and T all change is called **polytropic**.

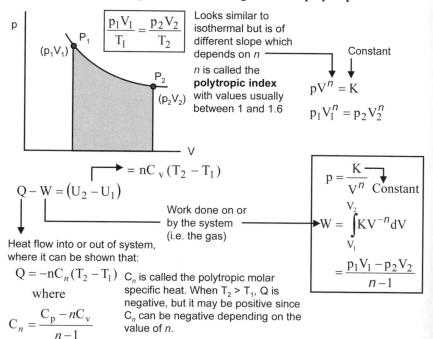

C_n is called the polytropic molar specific heat. When $T_2 > T_1$, Q is negative, but it may be positive since C_n can be negative depending on the value of n.

$$Q = -nC_n(T_2 - T_1)$$

where

$$C_n = \frac{C_p - nC_v}{n-1}$$

A gas process in which no heat flow occurs is called **adiabatic.**

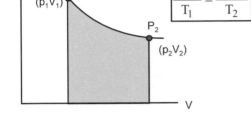

$Q = 0$

$$\boxed{\dfrac{p_1 V_1}{T_1} = \dfrac{p_2 V_2}{T_2}}$$

This is a very important process for heat engines since it represents the most efficient transfer of internal energy into work without loss to surroundings. Adiabatic processes can be approximated in systems that are insulated, have a small temperature difference to surroundings, or proceed very quickly such that heat has no time to flow into or out from system.

For adiabatic process, work done on or by the gas is equal to the change in internal energy.

$$= nC_v (T_2 - T_1)$$

$$Q - W = (U_2 - U_1)$$

$$W = \dfrac{p_1 V_1 - p_2 V_2}{\gamma - 1}$$

Heat flow into or out of system = 0

$$Q = -nC_n (T_2 - T_1)$$

$$= 0$$

$T_2 <> T_1$ because then the process would be isothermal.

$$\therefore 0 = nC_n$$

$$C_n = \dfrac{C_p - nC_v}{n - 1}$$

$$0 = \dfrac{n(C_p - nC_v)}{n - 1}$$

$$n = \dfrac{C_p}{C_v}$$

$$= \gamma \quad \text{Adiabatic index}$$

Adiabatic process is special case of the polytropic process with $n = \gamma$

$$\boxed{\begin{array}{l} pV^\gamma = K \\ p_1 V_1^\gamma = p_2 V_2^\gamma \end{array}}$$

Constant volume, constant pressure, isothermal and adiabatic processes are all special cases of the **polytropic** process.

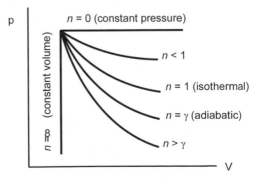

1.6.2 Example

1. 1 mole of nitrogen is maintained at atmospheric pressure and is
   heated from room temperature 20 °C to 100 °C:
   (a) Draw a pV diagram
   (b) Calculate the heat flow into the gas.
   (c) Calculate change in internal energy of the gas.
   (d) Calculate the work done by the gas.

Solution:

(a)

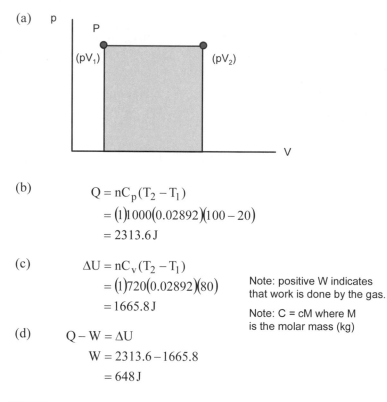

(b)
$$Q = nC_p(T_2 - T_1)$$
$$= (1)1000(0.02892)(100 - 20)$$
$$= 2313.6\,J$$

(c)
$$\Delta U = nC_v(T_2 - T_1)$$
$$= (1)720(0.02892)(80)$$
$$= 1665.8\,J$$

Note: positive W indicates
that work is done by the gas.

Note: C = cM where M
is the molar mass (kg)

(d)
$$Q - W = \Delta U$$
$$W = 2313.6 - 1665.8$$
$$= 648\,J$$

Nitrogen
c_p = 1000 J kg^{-1}K^{-1}
c_v = 720 J kg^{-1}K^{-1}
R = 8.3145 J mol^{-1} K^{-1}
Molecular weight = 28.92

1.7 Kinetic theory of gases

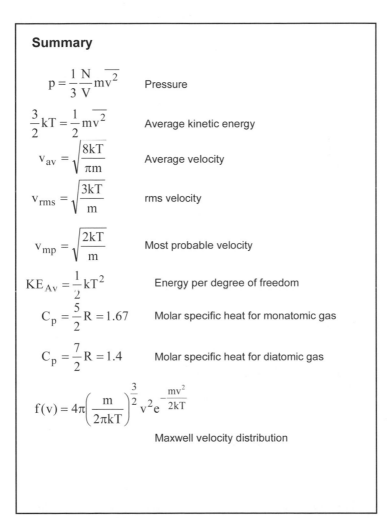

Summary

$$p = \frac{1}{3}\frac{N}{V}m\overline{v^2}$$ Pressure

$$\frac{3}{2}kT = \frac{1}{2}m\overline{v^2}$$ Average kinetic energy

$$v_{av} = \sqrt{\frac{8kT}{\pi m}}$$ Average velocity

$$v_{rms} = \sqrt{\frac{3kT}{m}}$$ rms velocity

$$v_{mp} = \sqrt{\frac{2kT}{m}}$$ Most probable velocity

$$KE_{Av} = \frac{1}{2}kT^2$$ Energy per degree of freedom

$$C_p = \frac{5}{2}R = 1.67$$ Molar specific heat for monatomic gas

$$C_p = \frac{7}{2}R = 1.4$$ Molar specific heat for diatomic gas

$$f(v) = 4\pi\left(\frac{m}{2\pi kT}\right)^{\frac{3}{2}}v^2 e^{-\frac{mv^2}{2kT}}$$

Maxwell velocity distribution

1.7.1 Pressure

Consider N molecules of an ideal gas inside a container of volume V at an absolute temperature T.

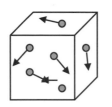

- The molecules are in rapid motion and move randomly around colliding with each other and the walls of the container.
- The molecules exert forces on the walls of the container during collisions
- During a collision, the velocity component v_y of the molecule is unchanged but v_x changes in direction but not in magnitude

Pressure is the result of the total average force acting on the walls of the container. Consider the collision of one molecule with the contain wall:

The change in velocity of a molecule during a time interval Δt is:

$$v_x - -v_x = 2v_x$$
$$\Delta v = 2v_x$$

Thus, the force imparted to the wall by the molecule is:

$$F = \frac{m\Delta v}{\Delta t} = \frac{m2v_x}{\Delta t}$$

$\Delta v/\Delta t$ is acceleration

During a time Δt, molecules a distance less than or equal to $v_x\Delta t$ away from the wall will strike the wall. Thus, the number of collisions will be the number of molecules within the volume element $\Delta V = Av_x\Delta t$.

If there are N molecules in the total volume V, then the number within the volume element ΔV is:

$$N\frac{\Delta V}{V}$$

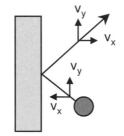

Total No. molecules

Half the molecules hit the wall, the other half are travelling in the other direction

Volume ΔV within which half the molecules hit the wall of area A

No. of collisions $= \dfrac{1}{2}N\dfrac{Av_x\Delta t}{V}$ ⟶ Volume of container

The total force on the wall at
any instant during time Δt is thus:

$$F_{total} = \left[\frac{m2v_x}{\Delta t}\right]\left[\frac{1}{2}\frac{N}{V}Av_x\Delta t\right]$$

Force from each collision — Total number of collisions

But, so far we have assumed that
v_x is the same for each molecule.
The molecules in the container

$$\frac{F}{A} = \frac{N}{V}m\overline{v_x^2}$$

actually have a range of speeds. The *average* value of v_x^2 components
leads to the **average force** (and hence pressure) on the wall:

$$\overline{v_x^2} = \frac{v_{x1}^2 + v_{x2}^2 + v_{x3}^2 + \dots}{N}$$

$$\frac{F_{av}}{A} = \frac{N}{V}m\overline{v_x^2} \quad = \text{Pressure}$$

But, it would be more convenient to have an expression which included the
total velocity v rather than the x component v_x, thus:

$$\overline{v^2} = \overline{v_x^2} + \overline{v_y^2} + \overline{v_z^2}$$
Magnitude of average velocity^2 given by sum of average components

$$\overline{v_x^2} = \frac{1}{3}\overline{v^2}$$ ← Since random motion in all directions thus velocity components are all equal

$$\overline{v_x^2} = \overline{v_y^2} = \overline{v_z^2}$$

$$\overline{v_x^2} = \frac{1}{3}\overline{v^2}$$

$$\frac{F_{av}}{A} = \frac{N}{V}m\overline{v_x^2}$$

$$\boxed{p = \frac{1}{3}\frac{N}{V}m\overline{v^2}}$$ → $v_{rms} = \sqrt{\overline{v^2}}$ The square root of the average of all the velocity^2 is called the **root mean square velocity**

$$pV = \frac{1}{3}Nm\overline{v^2}$$

$$= \frac{2}{3}N\left(\frac{1}{2}m\overline{v^2}\right)$$ Average translational kinetic energy of a single molecule

$$= nRT$$

→ But, R and N_A are constants. The ratio of them is a new constant, **Boltzmann's constant** k.

$$nRT = \frac{2}{3}N\left(\frac{1}{2}m\overline{v^2}\right)$$ Since $N = nN_A$

$$= \frac{2}{3}nN_A\left(\frac{1}{2}m\overline{v^2}\right)$$

$$\boxed{\frac{3}{2}kT = \frac{1}{2}m\overline{v^2}}$$ or $\frac{3}{2}kT = \frac{1}{2}mv_{rms}^2$

Average translational kinetic energy

1.38×10^{-23} J K^{-1}

$$\frac{3}{2}\left[\frac{R}{N_A}\right]T = \frac{1}{2}m\overline{v^2}$$

The **average translational kinetic energy** of a single molecule depends only on the temperature T.

1.7.2 Velocity distributions

Gas molecules in a container do not all have the same speed.

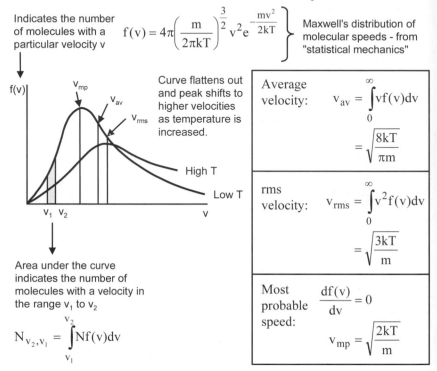

Indicates the number of molecules with a particular velocity v

$$f(v) = 4\pi\left(\frac{m}{2\pi kT}\right)^{\frac{3}{2}} v^2 e^{-\frac{mv^2}{2kT}}$$

Maxwell's distribution of molecular speeds - from "statistical mechanics"

Curve flattens out and peak shifts to higher velocities as temperature is increased.

Average velocity:

$$v_{av} = \int_0^\infty vf(v)dv$$

$$= \sqrt{\frac{8kT}{\pi m}}$$

rms velocity:

$$v_{rms} = \int_0^\infty v^2 f(v)dv$$

$$= \sqrt{\frac{3kT}{m}}$$

Area under the curve indicates the number of molecules with a velocity in the range v_1 to v_2

$$N_{v_2, v_1} = \int_{v_1}^{v_2} Nf(v)dv$$

Most probable speed:

$$\frac{df(v)}{dv} = 0$$

$$v_{mp} = \sqrt{\frac{2kT}{m}}$$

We saw previously that the **average kinetic energy** of a single gas molecule in a container could be calculated from the temperature of the gas. It is often difficult to decide which velocity to use in a given situation.

We should use **rms velocity** when dealing with the kinetic energy of the molecules. We should use **average velocity** when the process under consideration (e.g. flow through a pipe) is affected by the molecules' velocity.

To calculate the rms velocity, we square the velocities first, then divide by the total number of molecules, and then take the square root. This is different to finding the average velocity since the act of squaring the velocities first weights the final answer to those larger velocities in the velocity distribution. Thus, v_{rms} is a little large than v_{av}.

$$v_{rms} = \sqrt{\overline{v^2}}$$

$$= \sqrt{\frac{3kT}{m}}$$

1.7.3 Molecular motion

Molecules in a gas are capable of **independent motion**. Consider a diatomic gas molecule:

(a) the molecule itself can travel
as a whole from one place to
another

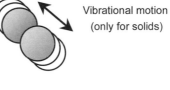

Translational motion

(b) the molecule can spin
around on its axis

Rotational motion

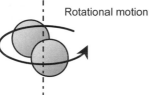

(c) the atoms within the molecule can
vibrate backwards and forwards

Vibrational motion
(only for solids)

Three translational velocity components are required to describe the motion of a **monatomic gas molecule** before and after any collisions. These components are called **degrees of freedom**.

1. v_z

Rotation about the axis of the atom is not counted since this does not change during collisions

2. v_x

3. v_y

Three **translational** and two **rotational** velocity components are required to describe the motion of a *diatomic* gas molecule before and after any collisions. This represents five degrees of freedom.

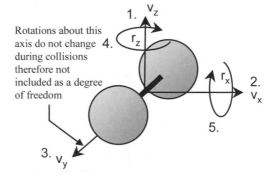

Rotations about this axis do not change during collisions therefore not included as a degree of freedom

1. v_z

4. r_z

2. v_x
r_x

3. v_y

5.

1.7.4 Specific heat and adiabatic index

The average kinetic energy for a gas molecule is equally partitioned between the rotational and translation components. For each degree of freedom, the average kinetic energy is given by $1/2\,kT$

Type of gas	Degrees of freedom	Total internal energy	
Monatomic gas	3	3/2 kT	⟶ all *translational* kinetic energy
Diatomic gas	5	5/2 kT ⎫	
Polyatomic gas	6*	6/2 kT ⎬ ⟶	Translational *and* rotational kinetic energy

* Could be more or less depending on the gas

Consider a temperature change $\Delta T = (T_2 - T_1)$ for n moles of a monatomic gas. For a given temperature rise, the change in average translational kinetic energy is equal to the change in internal energy. $\Delta U = nC_v\left(T_2 - T_1\right)$

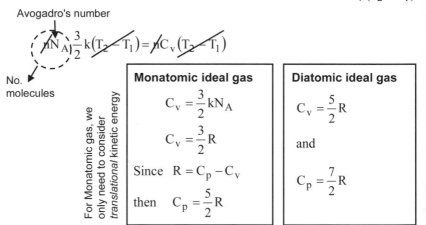

Avogadro's number

$$nN_A\frac{3}{2}k(T_2 - T_1) = nC_v(T_2 - T_1)$$

No. molecules

For Monatomic gas, we only need to consider *translational* kinetic energy

Monatomic ideal gas

$$C_v = \frac{3}{2}kN_A$$

$$C_v = \frac{3}{2}R$$

Since $R = C_p - C_v$

then $C_p = \frac{5}{2}R$

Diatomic ideal gas

$$C_v = \frac{5}{2}R$$

and

$$C_p = \frac{7}{2}R$$

The principle of equi-partition of energy allows us to calculate the **adiabatic index** of different types of gases.

For any ideal gas:

$$\gamma = \frac{C_p}{C_v}$$

Monatomic	Diatomic	Polyatomic
$\gamma = \frac{5}{2}R \cdot \frac{2}{3}\frac{1}{R}$	$\gamma = \frac{7}{2}R \cdot \frac{2}{5}\frac{1}{R}$	$\gamma = \frac{8}{2}R \cdot \frac{2}{6}\frac{1}{R}$
$= \frac{5}{3}$	$= \frac{7}{5}$	$= \frac{8}{6}$
$= 1.67$	$= 1.4$	$= 1.33$

1.7.5 Energy distributions

$$dN = Nf(v)dv$$

Shape of distribution curve depends on relative size of the translational kinetic energy and kT.

$$f(v) = 4\pi \left(\frac{m}{2\pi kT}\right)^{\frac{3}{2}} v^2 e^{-\frac{mv^2}{2kT}}$$

Maxwell *velocity* distribution function

Low value of kT means that the molecules tend to all have similar speeds (ordered motion). High value indicates a large range in speeds (disorder).

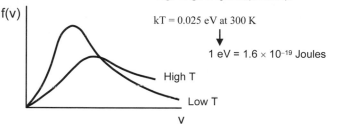

kT = 0.025 eV at 300 K

1 eV = 1.6 × 10⁻¹⁹ Joules

High T

Low T

The number of particles N with a total energy E was computed by Boltzmann using the **Maxwell velocity distribution** function:

$$N = Ce^{\frac{-E}{kT}}$$ Maxwell-Boltzmann *energy* distribution

a constant

and thus the ratio of numbers of molecules in a gas with energies E_1 and E_2 at a particular temperature T is given by:

$$\frac{N_1}{N_2} = \frac{e^{-E_1/kT}}{e^{-E_2/kT}}$$

1.7.6 Examples

1. Compute the average kinetic energy of a molecule of a gas at room temperature (300K):

Solution:

$k = 1.38 \times 10^{-23}$ J K^{-1}
$N_A = 6.02 \times 10^{23}$

$$KE_{av} = \frac{3}{2}kT$$

$$= \frac{3}{2}1.38 \times 10^{-23}300$$

$$= 6.2 \times 10^{-21} \text{ J}$$

2. If the gas in the previous question was H_2, calculate the rms velocity of a single molecule.

Solution:

$m_H = 3.32 \times 10^{-27}$kg

$$v_{rms} = \sqrt{\frac{3kT}{m}}$$

$$= \sqrt{\frac{3(1.38 \times 10^{-23})300}{3.32 \times 10^{-27}}}$$

$$= 1934 \text{ m s}^{-1}$$

3. For one mole of gas at 300K, calculate the total translational kinetic energy.

Solution:

$$KE_{Total} = N_A \frac{3}{2}kT$$

$$= 6.02 \times 10^{23} \frac{3}{2}1.38 \times 10^{-23}300$$

$$= 3750 \text{ J}$$

1.8 Heat engines

Summary

A heat engine is any device which is capable of continuous conversion of heat into work.

$$\eta_o = 1 - \frac{Q_R}{Q_S}$$ Carnot efficiency

$$= 1 - \frac{T_C}{T_H}$$

$$COP = \frac{T_C}{T_H - T_C}$$ Coefficient of performance

1.8.1 Cyclic processes

1. A gas undergoes a constant volume process from P_1 to P_2

2. It then undergoes an isothermal expansion to P_3

3. And finally a constant pressure compression back to P_1

All these are thermodynamic, quasi-static, **reversible processes**.

Quasi-static means that the pressure, temperature and volume are changing very slowly, or slowly enough so that dynamic effects (such as momentum changes or viscosity) are insignificant.

The word reversible means that the direction of the process may be reversed by simply reversing the temperature, pressure, volume, etc (or whatever is causing the state to change). An isothermal expansion may be reversed by reversing the direction of the work W in which case work W will be converted into heat Q and transferred back to the heat source.

Let us now combine these processes on the one diagram. This combination, where the pressure, temperature and volume of the **working fluid** have returned to their initial values, is a **thermodynamic cycle**.

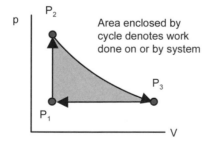

Area enclosed by cycle denotes work done on or by system

The cyclic nature of this arrangement means that continuous work output may be obtained. The continuous conversion of heat into work is the overriding characteristic of a **heat engine**. Other devices that convert heat into work in a non-cycle manner (e.g. a rifle that converts heat into kinetic energy of a bullet) is not a heat engine.

1.8.2 Otto cycle

The **Otto cycle** starts off with the first half of a **pumping cycle**, which is interrupted by the actual **working cycle**, after which the pumping cycle resumes. This cycle forms the basis of nearly all the world's motor vehicles.

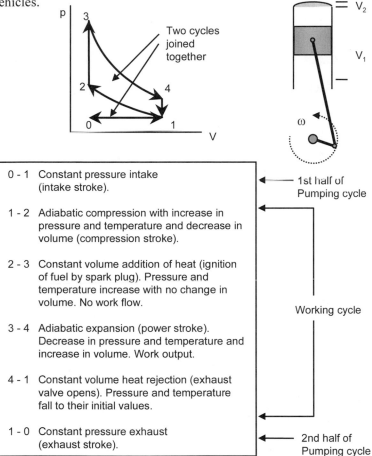

0 - 1   Constant pressure intake (intake stroke).

1 - 2   Adiabatic compression with increase in pressure and temperature and decrease in volume (compression stroke).

2 - 3   Constant volume addition of heat (ignition of fuel by spark plug). Pressure and temperature increase with no change in volume. No work flow.

3 - 4   Adiabatic expansion (power stroke). Decrease in pressure and temperature and increase in volume. Work output.

4 - 1   Constant volume heat rejection (exhaust valve opens). Pressure and temperature fall to their initial values.

1 - 0   Constant pressure exhaust (exhaust stroke).

The maximum theoretical (or indicated) thermal efficiency of an engine using the Otto cycle depends on the compression ratio V_1/V_2 and is given by:

$$\eta_i = 1 - \frac{1}{\left(\dfrac{V_1}{V_2}\right)^{\gamma-1}}$$

γ is the adiabatic index of the working gas.

1.8.3 Thermal efficiency

A **heat engine** uses a thermodynamic cycle to perform work. After a heat engine completes a cycle, the system returns to its original state. The only observable difference is that heat ΔQ_S has been taken from the source, a quantity of this heat has been converted to work ΔW, and the remainder ΔQ_R has been rejected into the **heat sink**.

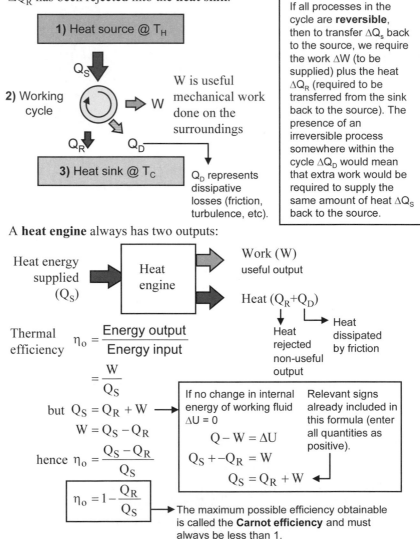

If all processes in the cycle are **reversible**, then to transfer ΔQ_S back to the source, we require the work ΔW (to be supplied) plus the heat ΔQ_R (required to be transferred from the sink back to the source). The presence of an irreversible process somewhere within the cycle ΔQ_D would mean that extra work would be required to supply the same amount of heat ΔQ_S back to the source.

Q_D represents dissipative losses (friction, turbulence, etc).

A **heat engine** always has two outputs:

Heat energy supplied (Q_S) → Heat engine → Work (W) useful output

Heat (Q_R+Q_D) → Heat rejected non-useful output → Heat dissipated by friction

Thermal efficiency

$$\eta_o = \frac{\text{Energy output}}{\text{Energy input}}$$

$$= \frac{W}{Q_S}$$

but $Q_S = Q_R + W$

$W = Q_S - Q_R$

hence $\eta_o = \dfrac{Q_S - Q_R}{Q_S}$

$$\boxed{\eta_o = 1 - \frac{Q_R}{Q_S}}$$

If no change in internal energy of working fluid $\Delta U = 0$

$$Q - W = \Delta U$$

$$Q_S + -Q_R = W$$

$$Q_S = Q_R + W$$

Relevant signs already included in this formula (enter all quantities as positive).

The maximum possible efficiency obtainable is called the **Carnot efficiency** and must always be less than 1.

1.8.4 Carnot cycle

Conditions for **maximum efficiency**:

- Heat is supplied isothermally at T_H
- Heat is rejected isothermally at T_C
- No heat is transferred anywhere else in the cycle

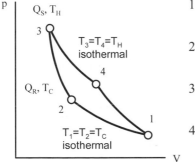

1 - 2 Isothermal compression
   Heat rejected into sink at T_C as pressure increased

2 - 3 Adiabatic compression
   Pressure and temperature both increase without any heat flow

3 - 4 Isothermal expansion
   Heat supplied from source at T_H and pressure decreases $p_1V_1 = p_2V_2$

4 - 1 Adiabatic expansion
   Pressure and temperature both decrease to initial value, no heat flow (i.e. lost to surroundings)

For $1-2$:
$$Q_R = W_{1-2} = p_1 V_1 \ln \frac{V_1}{V_2}$$

For $3-4$:
$$Q_S = W_{3-4} = p_3 V_3 \ln \frac{V_4}{V_3}$$

Now, $p_3 V_3 = nRT_3 \longrightarrow$ $T_3 = T_4 = T_H$ isothermal

$$Q_S = nRT_3 \ln \frac{V_4}{V_3}$$

$$= nRT_H \ln \frac{V_4}{V_3}$$

Also, $Q_R = nRT_C \ln \frac{V_1}{V_2}$

Since $\eta_c = 1 - \dfrac{Q_R}{Q_S}$

Then $\eta_c = 1 - \dfrac{T_C \ln V_1/V_2}{T_H \ln V_4/V_3}$

$$\boxed{\eta_c = 1 - \frac{T_C}{T_H}}$$

for adiabatic compression 2 - 3

$$\frac{T_3}{T_2} = \left(\frac{V_2}{V_3}\right)^{\gamma-1} = \frac{T_H}{T_C}$$

for adiabatic expansion 4 - 1

$$\frac{T_4}{T_1} = \left(\frac{V_1}{V_4}\right)^{\gamma-1} = \frac{T_H}{T_C}$$

thus

$$\frac{V_1}{V_2} = \frac{V_4}{V_3}$$

$$\boxed{\begin{array}{l} \dfrac{p_1V_1}{T_1} = \dfrac{p_2V_2}{T_2} \\[2mm] p_1V_1^{\gamma} = p_2V_2^{\gamma} \\[2mm] \dfrac{T_2}{T_1} = \left(\dfrac{V_1}{V_2}\right)^{\gamma-1} \end{array}}$$

Maximum possible efficiency attainable by any heat engine operating between temperature limits T_H and T_C

1.8.5 Heat sink in an engine

A quantity of heat is supplied from Q_S and work is done by piston against opposing force. If there is no friction, and the cylinder is perfectly insulating, then all the heat supplied Q_S goes into work.

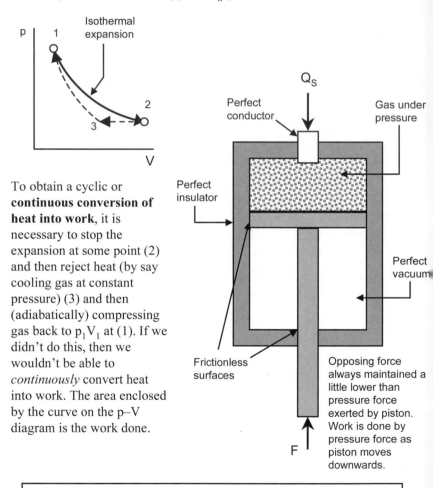

To obtain a cyclic or **continuous conversion of heat into work**, it is necessary to stop the expansion at some point (2) and then reject heat (by say cooling gas at constant pressure) (3) and then (adiabatically) compressing gas back to $p_1 V_1$ at (1). If we didn't do this, then we wouldn't be able to *continuously* convert heat into work. The area enclosed by the curve on the p–V diagram is the work done.

> *Continuous* or **cyclic conversion** of heat to work requires heat to be rejected at some point in the cycle.

usually to the surroundings

1.8.6 Reversibility

In a real heat engine, **irreversible processes** lead to dissipation of heat into the surroundings. Schematically, an engine can be thus treated as a series combination of a reversible cycle, the output of which is the work W_G done by the working substance, being then connected to an irreversible process, where some of the energy W_G is dissipated as heat Q_D, and the remainder available as useful work W_S.

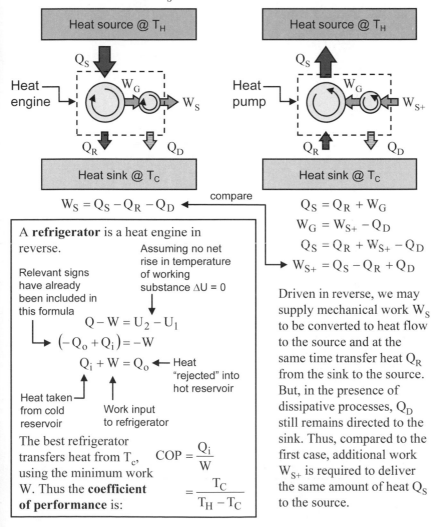

$$W_S = Q_S - Q_R - Q_D \xleftarrow{\text{compare}}$$

$$Q_S = Q_R + W_G$$
$$W_G = W_{S+} - Q_D$$
$$Q_S = Q_R + W_{S+} - Q_D$$
$$\longrightarrow W_{S+} = Q_S - Q_R + Q_D$$

A **refrigerator** is a heat engine in reverse.

Relevant signs have already been included in this formula

Assuming no net rise in temperature of working substance $\Delta U = 0$

$$Q - W = U_2 - U_1$$
$$\left(-Q_o + Q_i\right) = -W$$
$$Q_i + W = Q_o \longleftarrow \text{Heat "rejected" into hot reservoir}$$

Heat taken from cold reservoir

Work input to refrigerator

The best refrigerator transfers heat from T_c, using the minimum work W. Thus the **coefficient of performance** is:

$$COP = \frac{Q_i}{W}$$
$$= \frac{T_C}{T_H - T_C}$$

Driven in reverse, we may supply mechanical work W_S to be converted to heat flow to the source and at the same time transfer heat Q_R from the sink to the source. But, in the presence of dissipative processes, Q_D still remains directed to the sink. Thus, compared to the first case, additional work W_{S+} is required to deliver the same amount of heat Q_S to the source.

1.8.7 Examples

1. A **Carnot cycle** is operated between two heat reservoirs at 800 K and 300 K. If 600 J are withdrawn from the hot reservoir in each cycle, calculate the amount of heat rejected to the cold reservoir.

Solution:

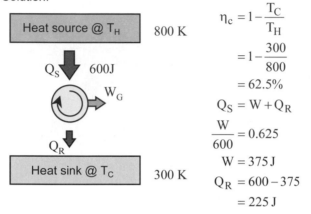

$$\eta_c = 1 - \frac{T_C}{T_H}$$

$$= 1 - \frac{300}{800}$$

$$= 62.5\%$$

$$Q_S = W + Q_R$$

$$\frac{W}{600} = 0.625$$

$$W = 375 \, J$$

$$Q_R = 600 - 375$$

$$= 225 \, J$$

2. The output from a petrol engine is 50 kW. The thermal efficiency is 15%. Calculate the heat supplied and the heat rejected in one minute.

Solution:

$$\frac{W}{60} = 50000$$

$$W = 3 \times 10^6 \, J$$

$$\eta = \frac{W}{Q_S} = 0.15$$

$$Q_s = 20 \, MJ$$

$$Q_s = Q_R + W$$

$$Q_R = 17 \, MJ$$

1.9 Entropy

Summary

Two most popular statements of the 2nd Law

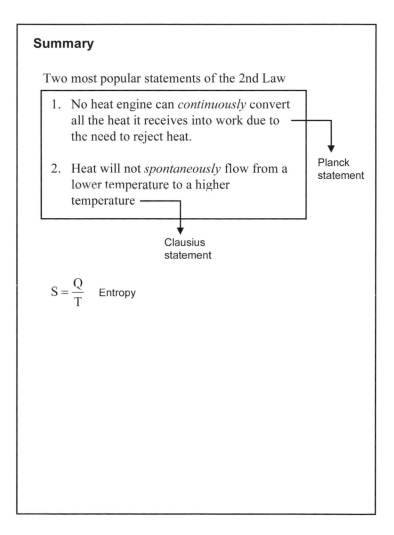

1. No heat engine can *continuously* convert all the heat it receives into work due to the need to reject heat.

2. Heat will not *spontaneously* flow from a lower temperature to a higher temperature

Planck statement

Clausius statement

$$S = \frac{Q}{T} \quad \text{Entropy}$$

1.9.1 Reversible and irreversible processes

Question:
A kilogram of water at 0 °C is mixed with 1 kilogram of water at 100 °C.
What happens?

$$\Delta Q = mc\Delta T$$
$$1(4186)(T_2 - 0) = -1(4186)(T_2 - 100)$$
$$T_2 = 50°C$$

$\longrightarrow$ We get 2 kg water at 50 °C

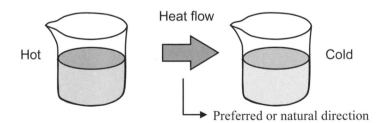

Hot   Cold

Warm

BUT WHY?

Certainly we can calculate that the heat lost by the hot water is equal to
the heat gained by the cold water, but why do we not just get 1 kg of hot
water and 1 kg of cold water in the same container. Energy would still be
conserved. *Why* does heat flow from the hotter to the colder body?
Why doesn't the water ever unmix itself into hot and cold regions?

Heat flow

Hot   Cold

$\longrightarrow$ Preferred or natural direction

The mixing of the water is an example of an **irreversible process**. Heat
flows from the hot to the cold water, no work is done. There is a preferred
or natural direction for all processes involving heat flow. The second law
is concerned with this **preferred direction**.

1.9.2 Entropy and reversibility

Consider a gas process:

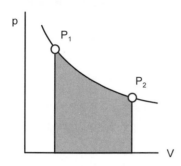

The area under the curve is the work done on or by the gas according to:

$$W = \int_{V_1}^{V_2} p(V)\,dV$$

The associated heat flow and internal energy change are given by the first law:

$$Q - W = (U_2 - U_1)$$

BUT! The first law says nothing about a very important experimental observation, and that is, what is the preferred or natural direction of the process? Is there a **natural direction** for this process anyway? Is it easier for the gas to expand and do work, or for us to do the work and compress the gas?

The word **reversible** means that the direction of the process may be reversed by simply reversing the temperature, pressure, volume, etc (or whatever is causing the state to change). There is no preferred or natural direction associated with work, it may be done on or by the gas with equal facility. But, there is a natural direction associated with heat flow. We shall see that heat flow to the surroundings leads to a process being irreversible. Work flow done by a system is reversible.

A measure of reversibility is called **entropy** and can be calculated using:

$$S = \frac{Q}{T}$$

Entropy

Note: Entropy is calculated from Q and T. Work does not come into it since there is no preferred or natural direction associated with work. There *is* a preferred or natural direction associated with heat and this is why entropy is calculated using Q.

S is a function of Q and T since there is a natural tendency for Q to travel from a hot body at T_2 to a cold body at T_1. That is, heat and temperature are the important variables.

1.9.3 Reversible process

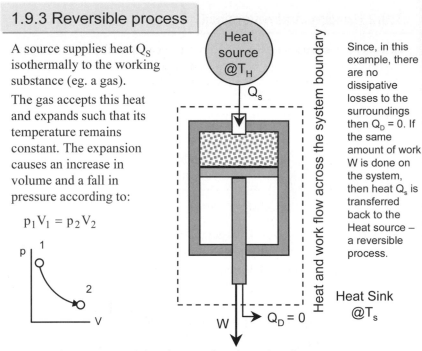

A source supplies heat Q_S isothermally to the working substance (eg. a gas).

The gas accepts this heat and expands such that its temperature remains constant. The expansion causes an increase in volume and a fall in pressure according to:

$$p_1 V_1 = p_2 V_2$$

Since, in this example, there are no dissipative losses to the surroundings then $Q_D = 0$. If the same amount of work W is done on the system, then heat Q_s is transferred back to the Heat source – a reversible process.

The pressure on the piston acting through a distance causes mechanical work W to be done. $W = P_1 V_1 \ln \dfrac{V_2}{V_1}$

Question: A reversible gas process happens within the system boundary. What is the entropy change <u>of the system</u> due to this event?

Answer:

In calculating changes in entropy, we need only consider those items or components in the system which accept or reject heat. Work transferred to surroundings does not affect entropy (there is no preferred direction for **work**).

Component	Heat flow
Working substance	$+ Q_S \ @ \ T_H$

Entropy change of the_**system**:

$$\Delta S = \frac{\Delta Q}{T}$$
$$= \frac{+Q_S}{T_H}$$

If this is a reversible process, then perhaps we might think that the change in entropy should be zero. We shall see why this is not so shortly.

1.9.4 Entropy change of the universe

So far, we have only considered the entropy change of the system. We now need to include the entropy changes of anything that absorbs or rejects hear in the vicinity of the system. Thus, for a **reversible process**, the entropy change of the system and things directly affected by the system is found from:

Component	Heat flow
Heat source	$-Q_S$ @ T_H
Working substance	$+Q_S$ @ T_H
Heat sink	$+Q_D = 0$ @ T_S

The system and the surroundings (heat sources and sinks) together may be referred to as the **universe**.

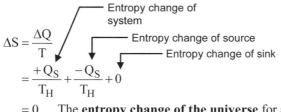

$$\Delta S = \frac{\Delta Q}{T}$$

Entropy change of system
Entropy change of source
Entropy change of sink

$$= \frac{+Q_S}{T_H} + \frac{-Q_S}{T_H} + 0$$

$= 0$ The **entropy change of the universe** for a reversible process is zero.

For an **irreversible process**, Q_d is not zero, hence, the entropy change of the system and things directly affected by the system is given by:

Component	Heat flow
Heat source	$-Q_S$ @ T_H
Working substance	$+Q_S$ @ T_H
Heat sink	$+Q_D$ @ T_S

Energy balance:

$$Q_S = W + Q_D$$
$$Q_S \neq W$$

Q_D does no work and hence can only be recovered (sent back to source) if additional work is supplied.

Heat flow to heat sink at T_S from dissipative mechanisms (e.g. friction)

Total entropy change:

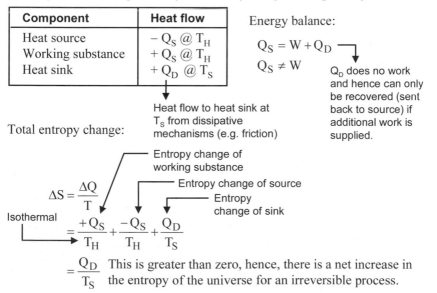

$$\Delta S = \frac{\Delta Q}{T}$$

Entropy change of working substance
Entropy change of source
Entropy change of sink

Isothermal

$$= \frac{+Q_S}{T_H} + \frac{-Q_S}{T_H} + \frac{Q_D}{T_S}$$

$$= \frac{Q_D}{T_S}$$ This is greater than zero, hence, there is a net increase in the entropy of the universe for an irreversible process.

1.9.5 Entropy in a cycle

Let us look at a heat engine utilising thermodynamic cycle and a working substance which consists of a series of reversible processes, e.g. a Carnot cycle

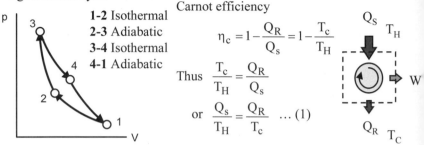

1-2 Isothermal
2-3 Adiabatic
3-4 Isothermal
4-1 Adiabatic

Carnot efficiency

$$\eta_c = 1 - \frac{Q_R}{Q_s} = 1 - \frac{T_c}{T_H}$$

Thus $\dfrac{T_c}{T_H} = \dfrac{Q_R}{Q_s}$

or $\dfrac{Q_s}{T_H} = \dfrac{Q_R}{T_c}$ ... (1)

Now, let us examine the change in entropy of the system by considering the heat gained and lost by the **working substance.** The system under consideration is just the working substance undergoing a series of processes.

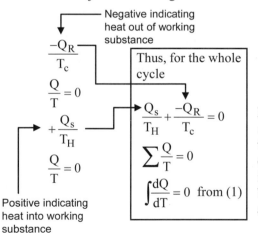

Negative indicating heat out of working substance

$$\frac{-Q_R}{T_c}$$

$$\frac{Q}{T} = 0$$

$$+\frac{Q_s}{T_H}$$

$$\frac{Q}{T} = 0$$

Positive indicating heat into working substance

Thus, for the whole cycle

$$\frac{Q_s}{T_H} + \frac{-Q_R}{T_c} = 0$$

$$\sum \frac{Q}{T} = 0$$

$$\int \frac{dQ}{dT} = 0 \text{ from (1)}$$

For any reversible *or irreversible* heat engine, there is no net change in entropy. This holds for irreversible processes because we are talking about a cycle.

A non-reversible process somewhere in the cycle represents a flow of heat to the surroundings Q_D at the expense of useful mechanical work available from the system.

If the heat engine itself is the system under consideration, then at the end of any cycle, the pressure, temperature, etc have all resumed their initial values. The fact that some heat Q_D has been generated at the expense of mechanical work does not affect the state *of the working substance* at the end of the cycle and there is no gain in entropy *within the engine.*

1.9.6 Entropy

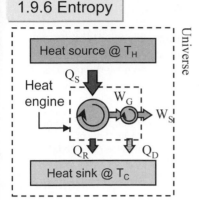

Note: It would appear that for a *reversible* engine ($Q_D = 0$), the fact that Q_R goes to the surroundings would lead to an overall increase in entropy since heat is being "lost" to the surroundings. However, this Q_R, even though it goes into the surroundings, is *recoverable* since it can be gotten back by supplying work W.

For the system and the surroundings (sources and sinks), there is no net change in entropy if all processes within the system are reversible. The total entropy increases if any processes within the system are **irreversible**.

- The heat source loses entropy $-Q_S/T_H$
- The heat sink gains entropy $+Q_R/T_C$
- The heat engine itself is returned to the same state at the end of each cycle so heat flow into and out of the working substance within the engine causes no net change in entropy within the engine.
- Work is done by the engine and may leave the system, but this does not lead to any change in entropy (work is **ordered energy**).
- For a reversible heat engine, $Q_S/T_H = Q_R/T_C$
- If there were a dissipative process, then less mechanical work would be performed and the heat sink would gain additional entropy $+Q_D/T_C$

Consider the mixing of hot and cold water. The temperature difference between the two may have been used as T_H and T_L for a heat engine. That is, we have the *opportunity* to do mechanical work. Once mixed and at uniform temperature, the opportunity to do mechanical work has been lost. Entropy is a measure of **lost opportunity**.

Entropy is a measure of **randomness** or **disorder**. Thermal energy arises due to random motion of molecules. Work flow, however, is ordered energy since it can be controlled. There is a natural tendency for things to become more disordered. Thus, any processing involving heat involves some natural tendency to disorder. This tendency towards disorder makes all real processes irreversible since the energy lost to the disordered state cannot be recovered unless additional work is done which is at the expense of additional disorder somewhere else in the universe. Entropy is a quantitative measure of the amount of disorder or randomness associated with a real process.

1.9.7 The 2nd Law of Thermodynamics

When a system undergoes a process, the entropy change of the system, added to the entropy change of the surrounding heat sources and sinks, is the total entropy change of the universe bought about by the process. The sign of ΔS (the total entropy change) signifies the presence of the following types of processes:

$\Delta S > 0$	irreversible process
$\Delta S = 0$	reversible process
$\Delta S < 0$	impossible process

Note: Entropy is not energy. There is no "Law of Conservation of Entropy". Indeed, irreversible processes "create" entropy.

All real processes involve some dissipative loss

All real processes are irreversible. The greater the irreversibility of a process, the greater the increase in entropy.

Another (but *quantitative*) statement of the 2nd Law

The **2nd law of thermodynamics** can be expressed in many ways. There is no single equation like the 1st law. There are three popular statements of the 2nd law:

1.	No heat engine can *continuously* convert heat into work due to the need to reject heat.	An engine which could convert *all* the heat it receives into work would spontaneously create order out of disorder. This has never been observed to happen.
2.	Heat will not *spontaneously* flow from a lower temperature to a higher temperature	There is a natural tendency towards the disordered state.
3.	The total entropy of any isolated system cannot decrease with time.	All real processes are irreversible which always leads to a total increase in entropy.

Energy has not been lost or destroyed, but the <u>opportunity to use it</u> has. The energy has become unavailable. Entropy is a quantitative measure of *lost opportunity*.

1.9.8 Examples

1. A kilogram of ice at 0 °C is
   melted and converted to water
   at 0 °C. What is the change in
   entropy of <u>this system</u>?

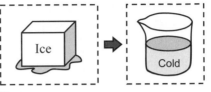

Solution:

Temperature is a constant at T = 273 K

Heat added = +335 kJ/kg (latent heat of fusion)

$$\Delta S = \int_{Q_1}^{Q_2} \frac{1}{T} dQ$$

$$= \frac{1}{T} \int_{Q_1}^{Q_2} dQ \quad \begin{array}{l} \text{T is a} \\ \text{constant} \end{array}$$

$$= \frac{335000}{273}$$

$$\boxed{= 1227 \text{ J K}^{-1}}$$ This is the entropy change of the system, not the universe.

Is this a reversible or irreversible process? We must consider the entropy change of the universe. If heat is transferred isothermally from some source, then the process is reversible since the entropy change of the source will be -1227 J K⁻¹. If heat is transferred from a source at a higher temperature, then the process is irreversible since Q_s/T_s will now be less than 1227 J K⁻¹.

2. A kilogram of water at 0 °C is
   heated to 100 °C. What is the
   change in entropy of <u>this system</u>?

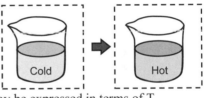

Solution:

Temperature is not a constant but Q may be expressed in terms of T.

$$dQ = mcdT$$

$$\Delta S = \int_{Q_1}^{Q_2} \frac{1}{T} dQ$$

Here we are approximating the process as an infinite series of isothermal processes taking in heat in infinitesimal quantities dQ.

$$= \int_{T_1}^{T_2} \frac{1}{T} mc \ dT$$

$$= mc \ln \frac{373}{273}$$ $c_{p \text{ water}}$ = 4186 J kg⁻¹K⁻¹

$$\boxed{= 1306 \text{ J K}^{-1}}$$ This is the entropy change of the system, not the universe.

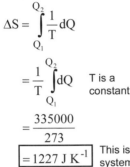

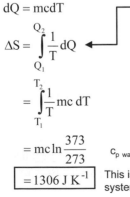

3. A kilogram of water at 0 °C is mixed with 1 kilogram of water at 100 °C.
    What is the *total* change in entropy?

Is this an irreversible process? Yes it is.
How can we tell? Well, in this process, the
only heat flow is from the hot water to the
cold water. There are no heat flows to and
from external sources and sinks. Hence,
the entropy change of this system is also
the entropy change of the universe. Since
there is a net increase in entropy (100J
K^{-1}), then the process must be irreversible.

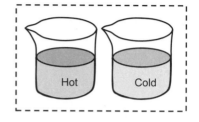

Solution: Take water at 0 °C to have **zero entropy**

Before mixing, total entropy is 1306 J K^{-1} + 0
After mixing, we get 2 kg water at 50 °C (323 K).
What is the entropy of the system after mixing?
Then can work out total ΔS.

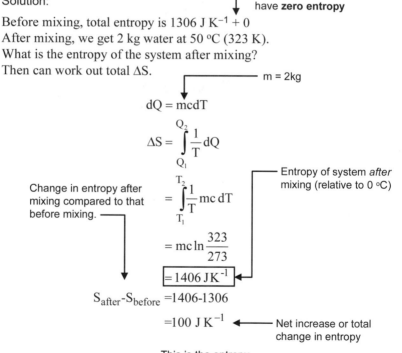

m = 2kg

$$dQ = mcdT$$

$$\Delta S = \int_{Q_1}^{Q_2} \frac{1}{T} dQ$$

Change in entropy after
mixing compared to that
before mixing.

Entropy of system *after*
mixing (relative to 0 °C)

$$= \int_{T_1}^{T_2} \frac{1}{T} mc\, dT$$

$$= mc \ln \frac{323}{273}$$

$$\boxed{= 1406\, J\,K^{-1}}$$

$$S_{after} - S_{before} = 1406 - 1306$$

$$= 100\ J\,K^{-1}$$

Net increase or total
change in entropy

This is the entropy
change of the universe.

Part 2

Waves
&
Optics

2.1 Periodic Motion

Summary

$y = A \sin \omega t$ — Displacement of particle from equilibrium position for Simple Harmonic Motion (SHM)

$y = A \sin(\omega t + \phi)$ — Displacement as a function of time with initial phase angle

$\dfrac{dy}{dt} = \omega A \cos(\omega t + \phi)$ — Velocity

$\dfrac{d^2 y}{dt^2} = -\omega^2 A \sin(\omega t + \phi)$ — Acceleration

$F = -ky$ — Restoring force characteristic of SHM

$T = 2\pi \sqrt{\dfrac{m}{k}}$ — Period of SHM

$E = \dfrac{1}{2} m \omega^2 A^2$ — Energy in SHM

$\omega = \sqrt{\dfrac{g}{L}}$ — Frequency of oscillation of a pendulum

2.1.1 Periodic motion

Consider the motion of a mass attached to a spring.

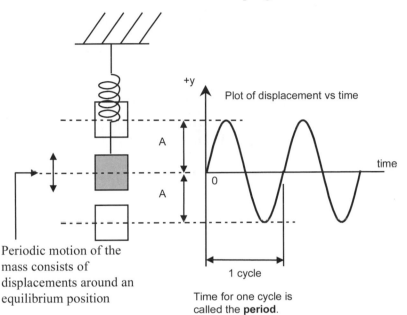

Plot of displacement vs time

time

1 cycle

Time for one cycle is called the **period**.

Periodic motion of the mass consists of displacements around an equilibrium position

Now, sin θ is a fraction which varies between 0 and ±1. If we multiply this fraction by the amplitude A of the motion then we get the displacement of the mass as a function of θ.

$$y = A \sin \theta$$

Since the mass undergoes **periodic motion** there is a **frequency** of oscillation. If the mass is moving up and down at a frequency ω, then the product ωt is the angle θ.

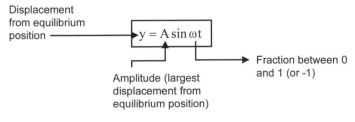

Displacement from equilibrium position

$y = A \sin \omega t$

Amplitude (largest displacement from equilibrium position)

Fraction between 0 and 1 (or -1)

2.1.2 Initial phase angle

Now, this expression: $y = A \sin \omega t$ assumes that $y = 0$ when $t = 0$
which is true if we start our time measurements from $y = 0$. What
happens if time measurements are started when the object is at $y = A$?

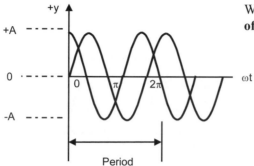

We have to add an initial
offset ϕ to the angle:

$$y = A \sin(\omega t + \phi)$$

initial phase angle

Now, at $t = 0$, we have: $y_0 = A \sin \phi$ where
y_0 is the initial displacement from the
equilibrium position. Consider these
examples of initial displacement y_0 for some
possible values of initial phase angles ϕ:

ϕ	y_0
$\pi/2$	A
π	0
$3/2\pi$	-A

The general expression for position (i.e. displacement from the equilibrium
position) for the mass is thus:

$$y = A \sin(\omega t + \phi)$$

Displacement

$$y = A \sin(\omega t + \phi)$$

Velocity

$$\frac{dy}{dt} = \omega A \cos(\omega t + \phi)$$

Acceleration

$$\frac{d^2 y}{dt^2} = -\omega^2 A \sin(\omega t + \phi)$$

2.1.3 Simple harmonic motion

Consider the force applied to a body moving with an oscillatory motion.

Now, $F = ma = -m\omega^2 A \sin(\omega t + \phi)$

but $y = A \sin(\omega t + \phi)$

thus $F = -m\omega^2 y$

However, the product $m\omega^2$ is actually a constant "k" for a fixed frequency:

Thus $\boxed{F = -ky}$ The minus sign indicates that the **force** acting on the mass is in a direction opposite to the displacement and acts so as to bring the mass back to the equilibrium position. The magnitude of this **restoring force** is a function of displacement of the mass from the equilibrium position. k is often called the spring or **force constant**.

The combination of periodic motion and a restoring force whose magnitude depends on the displacement from the equilibrium position is called **simple harmonic motion** or **SHM.**

Now, $\omega = \sqrt{\dfrac{k}{m}}$

also, $= 2\pi f = \dfrac{2\pi}{T}$

Because the displacements, velocities and accelerations of the mass involve smoothly varying sine and cosine functions.

$T = \dfrac{2\pi}{\omega}$

$\boxed{T = 2\pi\sqrt{\dfrac{m}{k}}}$ Note: the **period** of oscillation only depends on the mass and the spring (or force) constant and not on the amplitude.

In general, at any instant, the mass has **kinetic** and **potential** energy.

Potential

$P.E. = \int F\,dy = \int -ky\,dy$

$P.E. = \dfrac{1}{2}ky^2$

$= \dfrac{1}{2}kA^2 \sin^2(\omega t + \phi)$

Kinetic

$K.E. = \dfrac{1}{2}mv^2$

$= \dfrac{1}{2}m\omega^2 A^2 \cos^2(\omega t + \phi)$

As the sin term increases, the cos term decreases. In other words, for a constant m, ω and A, the total energy is a constant.

The **total energy** is P.E.+K.E., thus:

$E = \dfrac{1}{2}kA^2 \sin^2(\omega t + \phi) + \dfrac{1}{2}m\omega^2 A^2 \cos^2(\omega t + \phi)$

$= m\omega^2$

$E = \dfrac{1}{2}m\omega^2 A^2$

In SHM, energy is constantly being transferred between potential and kinetic.

2.1.4 Example

1. Determine an expression for the period of a pendulum.

L

θ

x

F

$mg\cos\theta$

mg

Equilibrium position

Solution:

Let x, the arc length, be the distance, or displacement, from the **equilibrium position**.

Whenever the pendulum is displaced from the equilibrium position, there is a **restoring force** F acting along x towards the equilibrium position.

$$F = -mg\sin\theta$$

For small angles of θ:

$$\sin\theta \approx \theta$$

Thus:

$$F = -mg\theta$$ → minus sign indicates force acts in opposite direction to positive x

$$= -mg\frac{x}{L}$$ → $x = L\theta$

$$= -\frac{mg}{L}x$$

$$k = \frac{mg}{L}$$ ⎤ this equation characteristic of SHM

$$F = -kx$$ ⎦

A pendulum does move in **simple harmonic motion** if the displacements from the equilibrium position are small.

Period of motion:

$$\omega = \sqrt{\frac{k}{m}}$$

$$= \sqrt{\frac{mg}{Lm}}$$

$$\omega = \sqrt{\frac{g}{L}}$$

In the case of a pendulum, the mass term cancels out and the **period** or **frequency** of motion depends only on the length of the pendulum and g.

2.2 Waves

Summary

$$v = f\lambda$$ Velocity of a wave

$$v = \sqrt{\frac{T}{\rho}}$$ Stretched string

$$y = A\sin(\omega t - kx + \phi)$$ Displacement of a particle

$$v_y = \omega A\cos(\omega t - kx + \phi)$$ Velocity of a particle

$$a_y = -\omega^2 A\sin(\omega t - kx + \phi)$$ Acceleration of a particle

$$\frac{\partial^2 y}{\partial x^2} = \frac{1}{v^2}\frac{\partial^2 y}{\partial t^2}$$ General wave equation

$$E = \frac{1}{2}\omega^2 A^2(\rho\lambda)$$ Energy transmitted by one wavelength of stretched string

$$P = \frac{1}{2}\sqrt{T\rho}\,\omega^2 A^2$$ Power in one wavelength on stretched string

2.2.1 Waves

A wave is some kind of a **disturbance**.
The disturbance travels from one place to another.
Some examples are:

- mechanical waves
- electromagnetic waves
- matter waves (probability waves)

Mechanical waves travel through a **medium** (eg. water). The **particles** which make up the medium are displaced from their equilibrium position in a regular periodic manner (which *may* be SHM - it depends on the system).

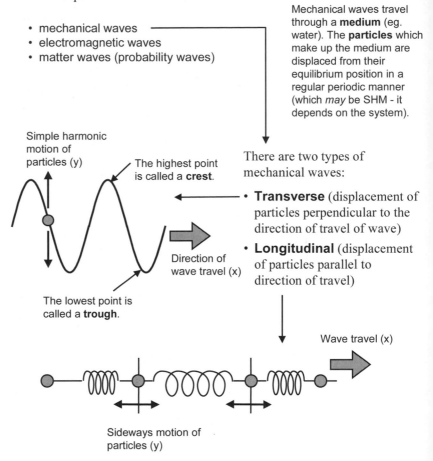

Simple harmonic motion of particles (y)

The highest point is called a **crest**.

Direction of wave travel (x)

The lowest point is called a **trough**.

There are two types of mechanical waves:

- **Transverse** (displacement of particles perpendicular to the direction of travel of wave)
- **Longitudinal** (displacement of particles parallel to direction of travel)

Wave travel (x)

Sideways motion of particles (y)

In a longitudinal wave, the displacements of the particles (y) is in the same direction as the motion of the wave (x). However, it is convenient to label the sideways movements as y to distinguish them from the wave travel x.

2.2.2 Wave motion

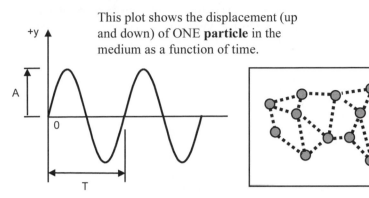

This plot shows the displacement (up and down) of ONE **particle** in the medium as a function of time.

Now, in a medium, there are many such particles. These particles (which might be **molecules** or **atoms** in a solid) are connected together by chemical bonds which act like springs.

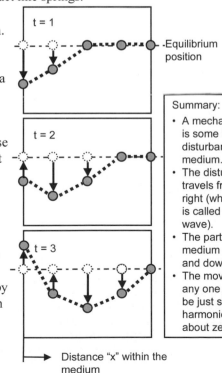

Let's look at a line of particles in the medium. If one particle is disturbed downwards, then this particle, after a short instant, will drag the next particle downwards. This particle, which of course is connected to the next particle, will drag that one down as well. But by this time, the first particle may be on the way back up. The particles all follow one another as they are dragged up and down by the previous particle. In this way, the **disturbance** or wave travels along through the medium.

t = 1

Equilibrium position

t = 2

t = 3

Distance "x" within the medium

Summary:
- A mechanical wave is some kind of disturbance in a medium.
- The disturbance travels from left to right (which is why it is called a travelling wave).
- The particles in the medium move up and down.
- The movement of any one particle may be just simple harmonic motion about zero.

2.2.3 Wavelength

At any particular instant, some particles in the medium are above their equilibrium position, some are below and some are at the equilibrium position. A **snapshot** of the particles in the medium looks like a wave:

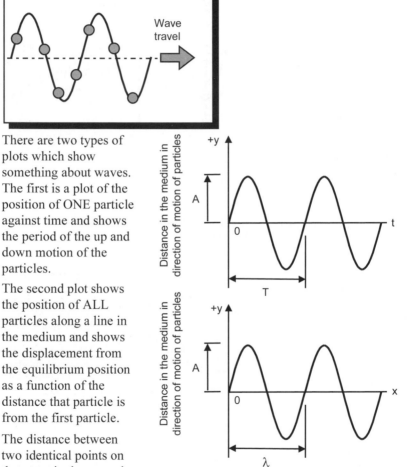

There are two types of plots which show something about waves. The first is a plot of the position of ONE particle against time and shows the period of the up and down motion of the particles.

The second plot shows the position of ALL particles along a line in the medium and shows the displacement from the equilibrium position as a function of the distance that particle is from the first particle.

The distance between two identical points on the wave in the second type of plot is called the **wavelength** λ.

These plots are different. The top one shows the displacement from equilibrium position against time for one particle. The bottom one shows the displacement from equilibrium position for a lot of particles at a particular instant in time.

2.2.4 Velocity, frequency and wavelength

Consider a **transverse wave** in a medium whose particles undergo SHM:

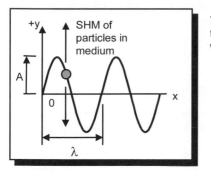

This is a "snapshot" of the disturbance, or wave, at some time "t".

• The shape of the wave is a repeating pattern.
  λ is called the **wavelength** and is the length of one complete cycle

The disturbance travels with a velocity v. The time for one complete cycle is T. Thus, since:

$$v = \frac{d}{t}$$ since one complete wavelength passes a given point in a time T

then $$v = \frac{\lambda}{T}$$ since $f = \frac{1}{T}$

and $$\boxed{v = f\lambda}$$ f in cycles per second

Action   **Observation**

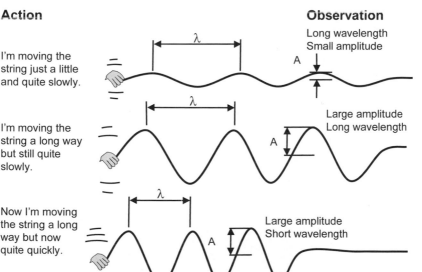

I'm moving the string just a little and quite slowly.

Long wavelength
Small amplitude

I'm moving the string a long way but still quite slowly.

Large amplitude
Long wavelength

Now I'm moving the string a long way but now quite quickly.

Large amplitude
Short wavelength

2.2.5 Wave velocity

It takes time for the motion of one atom or molecule to affect the next atom or molecule. This means that the original disturbance in the medium takes a finite amount of time to travel from one place to another. The **velocity of a wave** in a particular medium depends upon its physical properties (usually the density and elastic modulus).

The velocity of a wave is the speed of the disturbance as it passes through the medium - not the velocity of the particles within the medium.

Wave	Velocity	
Stretched string (transverse)	$v = \sqrt{\dfrac{T}{\rho}}$	T – tension in string ρ – mass per unit length
Fluid (longitudinal)	$v = \sqrt{\dfrac{B}{\rho}}$	B – bulk modulus ρ – density of fluid
Solid (longitudinal)	$v = \sqrt{\dfrac{E}{\rho}}$	E – elastic modulus ρ – density of solid
Gas (longitudinal)	$v = \sqrt{\dfrac{\gamma\, p}{\rho}}$ $= \sqrt{\dfrac{\gamma\, RT}{M}}$	γ – adiabatic index p – pressure T – abs. temp M – molar mass

Note that the velocity of the wave is a function of an elastic property/ inertial property

2.2.6 Particle displacement

Consider a transverse wave on a string. We wish to calculate the displacement y of any point P on the string as a function of time t. But, not all points on the string have the same displacement at any one time.

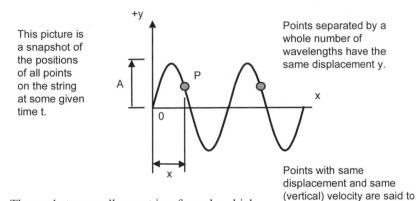

This picture is a snapshot of the positions of all points on the string at some given time t.

Points separated by a whole number of wavelengths have the same displacement y.

Points with same displacement and same (vertical) velocity are said to be **in phase** with each other.

Thus, what we really want is a formula which gives displacement y as a function of both x and t. If we had this formula, then we could find the displacement of a point P located at x at any time t.

$$y = f(x,t)$$

Let's consider the motion of the point located at x =0. If the points on the string are moving up and down with SHM, then:

$$y = A\sin(\omega t + \phi)$$ if y = 0 at
t = 0, then ϕ = 0

The disturbance, or wave, travels from left to right with velocity v = x/t. Thus, the disturbance travels from 0 to a point x in time x/v.

Now, let us consider the motion of a point P located at position x. The displacement of point P located at x at time t is the same as that of point P' located at x = 0 at time (t – x/v).

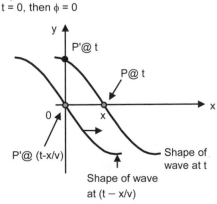

2.2.7 Wave equation

Thus, to get the displacement of the point P at (x,t) we use the same formula for the motion of point P' located at x = 0 but put in the time t = (t – x/v):

$$y = A\sin(\omega t + \phi)$$

$$= A\sin\left[\omega\left(t - \frac{x}{v}\right) + \phi\right]$$

Now, it is convenient to let: $k = \dfrac{2\pi}{\lambda}$

and since: $v = f\lambda$

then: $= f\dfrac{2\pi}{k}$ Wave number

$= \dfrac{\omega}{k}$

Thus: $y = A\sin\left[\omega\left(t - x\dfrac{k}{\omega}\right) + \phi\right]$

$$\boxed{y = A\sin(\omega t - kx + \phi)}$$ This applies to both transverse and longitudinal waves.

Displacement of particle in the medium from equilibrium position as a function of x and t.

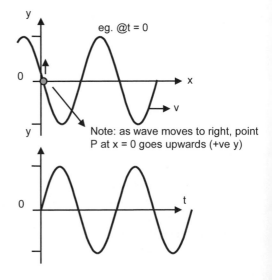

If t is held constant, and y plotted as a function of x, then shape of wave (snapshot) is displayed.

eg. @t = 0

Note: as wave moves to right, point P at x = 0 goes upwards (+ve y)

If x is held constant, and y plotted as a function of t, then SHM motion of single point is displayed.

2.2.8 General wave equation

The velocity and acceleration *of the particle* in the medium are found by differentiating while holding x constant:

Displacement

$$y = A\sin(\omega t - kx + \phi)$$

Velocity = dy/dt

$$v_y = \omega A\cos(\omega t - kx + \phi)$$ (holding x constant)

Acceleration = dv/dt

$$a_y = -\omega^2 A\sin(\omega t - kx + \phi)$$ (holding x constant)

Let us now find dy/dx while holding t constant:

$$y = A\sin(\omega t - kx)$$

$$\frac{\partial y}{\partial x} = -kA\cos(\omega t - kx)$$

$$\frac{\partial^2 y}{\partial x^2} = -k^2 A\sin(\omega t - kx)$$

The symbol ∂ is used to remind us that we are taking the derivative with one (or more) of the variables in the equation held constant (i.e. in this case, t). Derivatives of this type are called **partial derivatives**.

but

$$\omega = vk$$ v is the velocity of the *wave*

thus

$$\frac{\partial^2 y}{\partial x^2} = -\frac{\omega^2}{v^2} A\sin(\omega t - kx)$$

$$\frac{\partial^2 y}{\partial x^2} = \frac{1}{v^2}\frac{\partial^2 y}{\partial t^2}$$

This is called the **wave equation** and gives information about all aspects of the wave by tying together the motion of the particles and the wave.

a_y = acceleration of particle

Velocity of wave

2.2.9 Energy transfer by wave motion

A wave can be used to transfer energy between two locations.

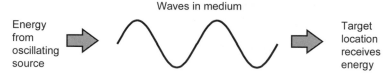

Waves in medium

Energy from oscillating source

Target location receives energy

1. The external source performs work on the first particle in the string.

2. The particle moves with SHM. Energy of particle is converted from P.E. to K.E. etc. Total energy of particle is unchanged.

3. Particle loses energy to next particle at the same rate it receives energy from the external source. Total energy of particle remains unchanged but energy from source gets passed on from one particle to the next till it arrives at the target location.

4. Energy from the external source travels along the string with velocity v.

5. The total energy of each particle is: $E = \frac{1}{2}m\omega^2 A^2$

6. The total energy for all oscillating particles in a segment of string one wavelength long is: $\boxed{E = \frac{1}{2}\omega^2 A^2 (\rho\lambda)}$ since $m = \rho\lambda$

↓ mass per unit length

In one time period T, the energy contained in one wavelength of string will have moved on to the next wavelength segment.

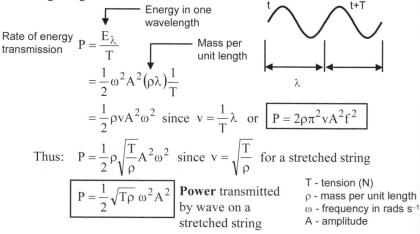

Energy in one wavelength

Rate of energy transmission $P = \frac{E\lambda}{T}$

Mass per unit length

$$= \frac{1}{2}\omega^2 A^2 (\rho\lambda)\frac{1}{T}$$

$$= \frac{1}{2}\rho v A^2 \omega^2 \text{ since } v = \frac{1}{T}\lambda \text{ or } \boxed{P = 2\rho\pi^2 v A^2 f^2}$$

Thus: $P = \frac{1}{2}\rho\sqrt{\frac{T}{\rho}}A^2\omega^2$ since $v = \sqrt{\frac{T}{\rho}}$ for a stretched string

$\boxed{P = \frac{1}{2}\sqrt{T\rho}\ \omega^2 A^2}$ **Power** transmitted by wave on a stretched string

T - tension (N)
ρ - mass per unit length
ω - frequency in rads s⁻¹
A - amplitude

2.2.10 Example

1. The following equation describes the displacement y (metres) of particles in a medium as a function of x and t:

$$y = 0.30\sin(151.4t - 8x + 0.75\pi)$$

(a) Calculate the amplitude of this wave?
(b) Calculate the frequency (in Hertz)?
(c) Calculate the velocity of the wave?
(d) What is the particle's velocity at a position x = 200 mm and t = 6 secs?

Solution:

$A = 0.30$m

$\omega = 151.4 \text{ rad s}^{-1}$

$f = \dfrac{151.4}{2\pi} = 24.1 \text{Hz}$

$v = f\lambda; k = \dfrac{2\pi}{\lambda}$

$+8 = \dfrac{2\pi}{\lambda}$

$v = (31.83)\left(\dfrac{2\pi}{8}\right) = 25 \text{ m s}^{-1}$

$v_y = \omega A \cos(\omega t - kx + \phi)$

$= (151.4)(0.3)\cos(151.4(6) - 8(0.20) + 0.75\pi)$

$= -14.9 \text{ m s}^{-1}$

> All these angles are in radians, so make sure your calculator is in radians mode before taking the cos!

Sign conventions:

Now, in this book, and in some other text books, the wave function is written:

$$y = A\sin(\omega t - kx + \phi) \quad (1)$$

But, in some books you will see:

$$y = A\sin(kx - \omega t + \phi) \quad (2)$$

Both are correct, but it depends on which direction is defined as being +ve for the vertical displacements of the particle. In this book, +ve y has meant upwards. Hence, in the figure here, the wave shape is drawn initially going *down* at t = 0 because as the wave moves to the right, the particle at x = 0 actually moves up (+ve).

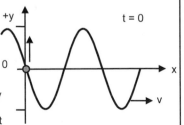

As wave moves to right, particle at x = 0 moves upwards.

However, some books show waves as initially going "up". In this case, the direction of +ve y is actually downwards. The formulas give different answers to problems. To get the same answer, $\phi = \pi$ may be used in the first equation. In this book, the +ve direction for y is upwards and hence formula (1) above will be used.

2.3 Superposition

Summary

$$y_R = 2A\sin(\omega t - kx + \phi)$$

Superposition of two waves of same amplitude, phase and direction

$$y_1 + y_2 = 2A\left[\cos\frac{2\pi x}{\lambda}\right]\sin\omega t$$

Superposition of two waves of same amplitude, phase but opposite direction

Fourier analysis

$$y = \left[\frac{4}{\pi}\sin\omega t + \frac{4}{3\pi}\sin 3\omega t + \frac{4}{5\pi}\sin 5\omega t + ...\right]$$

$$f = \frac{n}{2L}\sqrt{\frac{T}{\rho}}$$

Superposition of two waves of same amplitude, phase but opposite direction

$$2\frac{\omega_1 - \omega_2}{2} = \omega_1 - \omega_2$$

Beat frequency

$$\Delta p = v^2 \rho k A \cos(\omega t - kx)$$

Pressure amplitude

$$I = 2\pi^2 A^2 f^2 \rho v$$

Intensity

$$\beta = 10\log_{10}\frac{I}{I_o}$$

Decibels

$$c = \frac{1}{\sqrt{\mu_o \varepsilon_o}}$$

$$= 3\times 10^8 \, ms^{-1}$$

Speed of light

2.3.1 Superposition

What happens when two or more
waves arrive at the same point at the
same time?

The resulting displacement
of particles within the
medium is the sum of the
displacements that would
occur from each wave if it
were acting alone.

This is the PRINCIPLE OF **SUPERPOSITION**

The general wave equation ties together information about the shape of the
wave, and velocity and accelerations of the particles within the medium
through which the wave is travelling.

$$\frac{\partial^2 y}{\partial x^2} = \frac{1}{v^2} \frac{\partial^2 y}{\partial t^2}$$ General wave
equation

This leads to a more formal definition
of superposition.

If two wave functions $y_1(x,t)$ and $y_2(x,t)$ both satisfy the general wave
equation, then so does the resulting combined function y_1+y_2.

2.3.2 Interference

Interference is the term used to describe *the result* of the superposition of waves.

Constructive interference $\longrightarrow$ Amplitude of resultant is larger than either component

Destructive interference $\longrightarrow$ Amplitude of resultant is less than one of the components.

Consider two waves travelling in the same direction having the same amplitude, frequency and wavelength.

$$y_1 = A\sin(\omega t - kx + \phi_1)$$
$$y_2 = A\sin(\omega t - kx + \phi_2)$$

Case 1: Both have the same initial phase angle ϕ

$$\begin{aligned} y_R &= y_1 + y_2 \\ &= A\sin(\omega t - kx + \phi) + A\sin(\omega t - kx + \phi) \\ &= 2A\sin(\omega t - kx + \phi) \end{aligned}$$

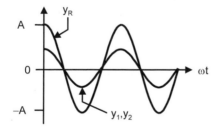

Resultant wave has twice the amplitude but the same frequency as the component waves.

Case 2: One wave has a phase angle $\phi = \pi$ and the other $\phi = 0$

$$y_1 = A\sin(\omega t - kx)$$
$$y_2 = A\sin(\omega t - kx + \pi)$$
$$y_1 + y_2 = A\sin(\omega t - kx) + A\sin(\omega t - kx + \pi)$$
$$\text{let } \theta = \omega t - kx$$
$$= A\sin(\theta) + A\sin(\theta + \pi)$$
$$\text{but } \sin(\theta + \pi) = -\sin\theta$$
$$y_1 + y_2 = 0 \quad \text{Waves cancel out.}$$

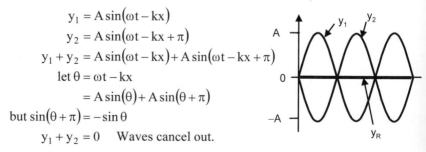

Case 3. Waves have a phase difference of $\phi = \pi/4$

$$y_1 = A\sin(\omega t - kx)$$
$$y_2 = A\sin(\omega t - kx + \pi/4)$$
$$y_1 + y_2 = A\sin(\omega t - kx) + A\sin(\omega t - kx + \pi/4)$$

but $\sin A + \sin B = 2\sin\left(\dfrac{A+B}{2}\right)\cos\left(\dfrac{A-B}{2}\right)$

thus $\quad y_1 + y_2 = A\left(2\sin\dfrac{(\omega t - kx) + (\omega t - kx + \pi/4)}{2}\right.$

$$\left.\cos\dfrac{(\omega t - kx) - (\omega t - kx + \pi/4)}{2}\right)$$

$$= \underbrace{2A\cos\dfrac{\pi}{8}}\sin(\omega t - kx + \pi/8)$$

Amplitude of resultant wave

Resultant wave is out of phase by $\pi/8$ to wave 1

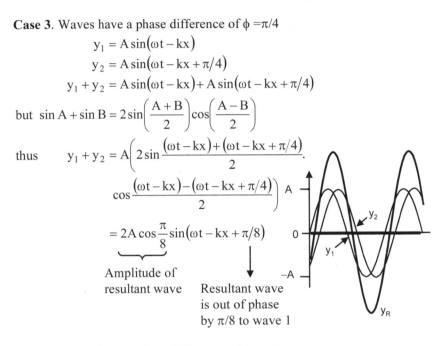

Case 4: Waves have a phase difference of $\phi = \pi/3$

$$y_1 = A\sin(\omega t - kx)$$
$$y_2 = A\sin(\omega t - kx + \pi/3)$$

$$y_1 + y_2 = A\left(2\sin\dfrac{(\omega t - kx) + (\omega t - kx + \pi/3)}{2}\right.$$

$$\left.\cos\dfrac{(\omega t - kx) - (\omega t - kx + \pi/3)}{2}\right)$$

$$= \underbrace{2A\cos\dfrac{\pi}{6}}\sin(\omega t - kx + \pi/6)$$

Amplitude of resultant wave

Resultant wave is out of phase by $\pi/6$ to wave 1

2.3.3 Fourier analysis

GENERAL CASE: frequency, amplitude and phase are all different.

Any periodic waveform, no matter how complicated, can be constructed by the superposition of sine waves of the appropriate frequency and amplitude. The component frequencies can be found by a mathematical technique called **Fourier analysis**.

Consider a square wave. The complete Fourier series for this is:

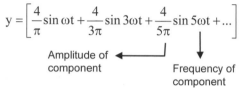

$$y = \left[\frac{4}{\pi}\sin \omega t + \frac{4}{3\pi}\sin 3\omega t + \frac{4}{5\pi}\sin 5\omega t + ...\right]$$

A **Fourier transform** of a wave is a "frequency map", it tells us what component waveforms are needed to produce the overall wave.

Amplitude of component ⟵⎯⎯⎯⎯⎯⎯⎯⎯⎯

Frequency of component

This means that a square wave can be obtained by superimposing a series of sine functions each of a different amplitude and frequency.

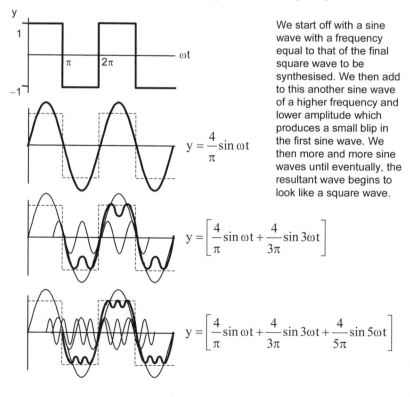

We start off with a sine wave with a frequency equal to that of the final square wave to be synthesised. We then add to this another sine wave of a higher frequency and lower amplitude which produces a small blip in the first sine wave. We then more and more sine waves until eventually, the resultant wave begins to look like a square wave.

$$y = \frac{4}{\pi}\sin \omega t$$

$$y = \left[\frac{4}{\pi}\sin \omega t + \frac{4}{3\pi}\sin 3\omega t\right]$$

$$y = \left[\frac{4}{\pi}\sin \omega t + \frac{4}{3\pi}\sin 3\omega t + \frac{4}{5\pi}\sin 5\omega t\right]$$

2.3.4 Superposition for waves in opposite directions

Let the waves have the same amplitude, wavelength and frequency but be travelling in opposite directions.

Wave travelling to the right:

$$y_1 = A\sin(\omega t - kx)$$

Wave travelling to the left:

$$y_2 = A\sin(\omega t + kx)$$

Resultant wave:

$$y_1 = A\sin(\omega t - kx)$$
$$y_2 = A\sin(\omega t + kx)$$

$$y_1 + y_2 = A\left(2\sin\frac{(\omega t - kx)+(\omega t + kx)}{2}\right.$$

$$\left.\cos\frac{(\omega t - kx)-(\omega t + kx)}{2}\right)$$

$$= 2A\cos kx \sin\omega t \quad\text{since } k = \frac{2\pi}{\lambda};\ \sin A + \sin B =$$

$$y_1 + y_2 = 2A\underbrace{\left[\cos\frac{2\pi x}{\lambda}\right]}\sin\omega t \qquad 2\sin\left(\frac{A+B}{2}\right)\cos\left(\frac{A-B}{2}\right)$$

Amplitude varies between 0 and 2A as x varies. The motion is still simple harmonic motion but the amplitude varies according to the value of x.

The resulting displacement is always zero when: $\dfrac{2\pi x}{\lambda} = n\dfrac{\pi}{2}$ $n = 1,3,5\ldots$

no matter what value of t.

These values of x are called **nodes** and occur when: $x = \dfrac{\lambda}{4}, \dfrac{3\lambda}{4}, \dfrac{5\lambda}{4}\ldots$

The resultant displacements are a maximum when: $\dfrac{2\pi x}{\lambda} = m\pi$ $m = 0,1,2,\ldots$

These positions are called **antinodes** and occur when: $x = 0, \dfrac{\lambda}{2}, \lambda, \dfrac{3\lambda}{2}\ldots$

2.3.5 Standing waves

All particles in the medium undergo SHM. It just so happens that the particles at the nodes always have an amplitude of zero and hence are stationary. The other particles usually oscillate very rapidly and hence the eye only sees the envelope of the waveform, hence the term **standing wave**.

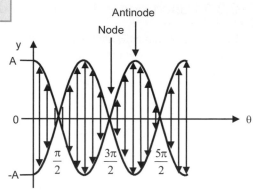

A very common example of standing waves appears in a stretched string. If the string is fixed at both ends then there must be at least a node at each end - because y = 0 at both ends.

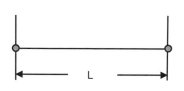

For a standing wave to be produced, the length of the string must be equal to an integral number of half-wavelengths.

$$L = n\frac{\lambda}{2}$$

Because only then is there at node at each end

$$n = 1,2,3,...$$

n = 1   1st harmonic

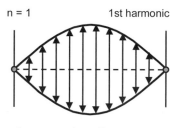

L = λ/2

The frequency of this mode of vibration is called the fundamental frequency.

Now, $v = f\lambda$

$$f = \frac{v}{\lambda}$$

$$v = \sqrt{\frac{T}{\rho}}$$ For a stretched string

$$f = \frac{1}{\lambda}\sqrt{\frac{T}{\rho}}$$

n = 2   2nd harmonic

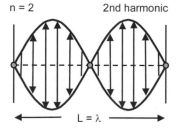

L = λ

But $L = \frac{n\lambda}{2}$

Thus $f = \frac{n}{2L}\sqrt{\frac{T}{\rho}}$

Allowable frequencies for standing waves (or normal modes) for a string length L.

2.3.6 Resonance

Consider a travelling wave on a stretched string:

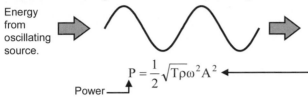

Energy
from
oscillating
source.

Target location
receives energy
and converts it
into some other
form.

$$P = \frac{1}{2}\sqrt{T\rho}\,\omega^2 A^2$$

Power

Now consider a standing wave on a stretched string:

n = 2

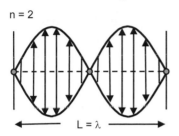

L = λ

If there were no energy
losses (eg. friction, sound)
then waves would travel
backwards and forwards
being reflected from each
end indefinitely.

If we kept adding energy to the system, then since there is no
dissipation, the amplitude of the waves would increase indefinitely.

The amplitude increases because the frequency (via the wavelength
and velocity) is fixed by the length of the string. The only thing that
may increase is A.

This is called **resonance** and occurs when:

- there are no energy dissipative mechanisms
- the oscillations are supplied at a frequency at
  or near a **normal mode** frequency.

If the oscillations are not provided at or near a normal mode (or **resonant
frequency**) then energy is not transferred into the system as effectively.

2.3.7 Beats

Consider two waves travelling in the same direction, same A, but with different frequencies.

$$y_1 = A\sin(\omega_1 t - kx)$$
$$y_2 = A\sin(\omega_2 t - kx)$$

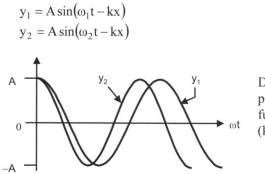

Displacement plotted as a function of time (keeping x fixed).

Resultant:

$$\sin A + \sin B =$$

$$2\sin\left(\frac{A+B}{2}\right)\cos\left(\frac{A-B}{2}\right)$$

$$\begin{aligned}
y_R &= y_1 + y_2 \\
&= A\sin(\omega_1 t - kx) + A\sin(\omega_2 t - kx) \\
&= 2A\cos\left[\frac{\omega_1 - \omega_2}{2}t\right]\sin\left[\frac{\omega_1 + \omega_2}{2}t - kx\right]
\end{aligned}$$

using

Amplitude term oscillates with time at a frequency of:

$$\frac{\omega_1 - \omega_2}{2}$$

Frequency of resultant oscillation is the average of ω_1 and ω_2

Amplitude of y_R oscillates with a frequency $(\omega_1 - \omega_1)/2$ but the ear hears two pulses or **beats** in this one cycle.

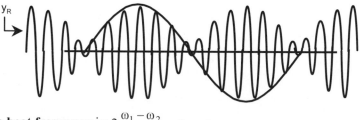

The **beat frequency** is: $2\dfrac{\omega_1 - \omega_2}{2} = \omega_1 - \omega_2$

2.3.8 Longitudinal waves

Consider some particles connected by series of springs as shown:

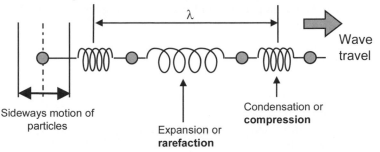

Sideways motion of particles

Expansion or **rarefaction**

Condensation or **compression**

Wave travel

As the particle moves to the right, spring is compressed and this compression acts on the next particle and so on. If the first particle moves with a regular period motion backwards and forwards about its equilibrium position, then eventually all the particles will move with this same oscillatory motion. If the first particle moves with SHM, then all all particles will move with SHM but with a phase lag compared to the first particle.

In a longitudinal wave, particles undergo oscillations in a direction parallel to the direction of the wave propagation. Longitudinal waves are really just like transverse waves except that the particles in the medium move back and forth instead of up and down.

Examples of longitudinal waves:

• Sound waves

• Waves on a slinky spring

• Earthquake waves (primary waves - the secondary earthquake waves are transverse).

• Any waves in a liquid or a gas (liquids and gases cannot transmit transverse waves - except on the surface of liquids)

The wave equation is exactly the same as before except that y indicates longitudinal displacements of particles from their equilibrium positions:

$$y = A\sin(\omega t - kx)$$

2.3.9 Sound waves

Longitudinal waves are alternate compressions and expansions between particles in the medium. For **sound waves**, the compression corresponds to pressure changes ≈ 1 Pa in the medium.

Pressure changes result in volume changes within the medium. The **bulk modulus** B is a measure of the relationship between the two.

$$B = -\frac{\Delta p}{\Delta V / V_o}$$

For longitudinal waves, the volume changes occur in a direction parallel to the displacements from equilibrium positions and the fractional volume change is proportional to $\Delta y / \Delta x$ or $\delta y / \delta x$.

$$\Delta p = -B \frac{\Delta V}{V_o}$$

$$= -B \frac{\Delta y}{\Delta x} = -B \frac{\partial y}{\partial x}$$

Hence, for sound waves in a solid:

$$\frac{\partial y}{\partial x} = kA \cos(\omega t - kx)$$

$$v = \sqrt{\frac{B}{\rho}} \quad \text{Velocity of wave}$$

$$B = v^2 \rho$$

$$\boxed{\Delta p = v^2 \rho kA \cos(\omega t - kx)}$$

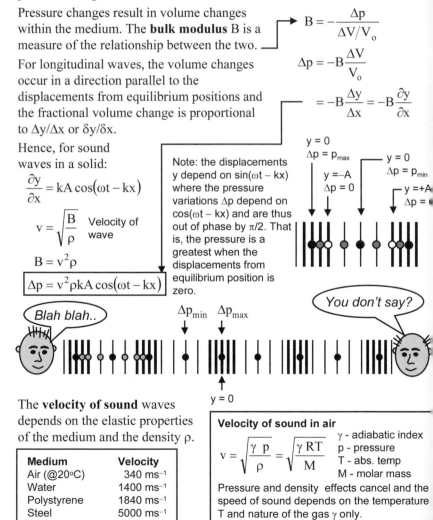

Note: the displacements y depend on $\sin(\omega t - kx)$ where the pressure variations Δp depend on $\cos(\omega t - kx)$ and are thus out of phase by $\pi/2$. That is, the pressure is a greatest when the displacements from equilibrium position is zero.

The **velocity of sound** waves depends on the elastic properties of the medium and the density ρ.

Medium	Velocity
Air (@20°C)	340 ms⁻¹
Water	1400 ms⁻¹
Polystyrene	1840 ms⁻¹
Steel	5000 ms⁻¹

Velocity of sound in air

$$v = \sqrt{\frac{\gamma\, p}{\rho}} = \sqrt{\frac{\gamma\, RT}{M}}$$

γ - adiabatic index
p - pressure
T - abs. temp
M - molar mass

Pressure and density effects cancel and the speed of sound depends on the temperature T and nature of the gas γ only.

Although the velocity of sound generally decreases with increasing density of the medium, it also depends on the medium's elastic properties. Thus, the velocity of sound in solids is greater than that in gases.

2.3.10 Superposition of longitudinal waves

Two longitudinal waves interfere constructively when the density or
pressure is enhanced by the superposition. Let us consider sounds waves in
a pipe where the waves have identical amplitude and frequency and are
travelling in opposite directions.

Case 1. Pipe is closed at
          one end.

Now, at the closed end of
the pipe, the displacements
of particles within the air
inside the pipe must be
zero since the closed end of
the pipe prevents any
displacements.

The allowable
frequencies for standing
waves are thus fixed by
the length of the pipe.

$$L = n\frac{\lambda}{4} \quad n = 1, 3, 5...$$

Odd harmonics of
the fundamental.

Plot of displacement y
of particles within the
air along the length of
the pipe for resultant
wave. +ve
displacements of
particles vary between
−A and +A.

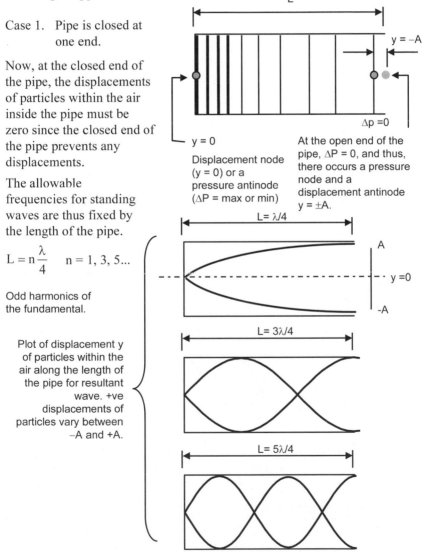

y = 0

Displacement node
(y = 0) or a
pressure antinode
(ΔP = max or min)

At the open end of the
pipe, ΔP = 0, and thus,
there occurs a pressure
node and a
displacement antinode
y = ±A.

y = −A

Δp = 0

L = λ/4

A

y = 0

−A

L = 3λ/4

L = 5λ/4

L

Case 2. Pipe is open at both ends.

At the open ends of the pipe, $\Delta P = 0$, and thus, there is a pressure node and a displacement anti-node $y = \pm A$. The allowable frequencies for standing waves are again fixed by the length of the pipe.

$$L = n\frac{\lambda}{2} \quad n = 1, 2, 3...$$

All harmonics of the fundamental.

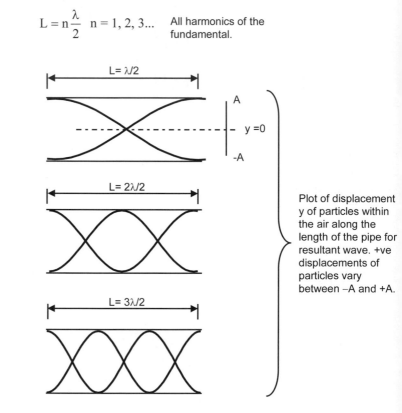

Plot of displacement y of particles within the air along the length of the pipe for resultant wave. +ve displacements of particles vary between −A and +A.

The diameter of the pipe makes a difference to the volume of the sound.

2.3.11 Sound intensity

Sound waves are longitudinal waves characterised by pressure (or density) variations. The power for a 1 m² square area of wave is given by:

$$\frac{P}{Area} = \frac{1}{2}\rho v A^2 \omega^2 \quad W\ m^{-2}$$

ρ in this formula is the mass per unit volume (or density) of the medium. This is similar to the expression given for the power transmitted on a string, but here, we have a 3 dimensional situation.

$$= \frac{1}{2}\rho v A^2 v^2 k^2$$

$$\frac{P}{Area} = \frac{\Delta p_{max}^2}{\rho v} \quad \text{since} \quad \Delta p_{max} = v^2 \rho k A$$

$$= I \quad \text{Intensity} \qquad\qquad \text{Amplitude}$$

or

$$\boxed{I = 2\pi^2 A^2 f^2 \rho v}$$

since $\omega = 2\pi f$

The pressure variations in sound are in the order of 1 Pa. The maximum in the pressure difference Δp_{max} is called the **pressure amplitude**.

$$\Delta p_{max} = v^2 \rho k A$$
$$= BkA$$

Linear and log scales

When something increases linearly, there is a steady increase in the quantity. When something increases logarithmically, there is a rapid increase in the quantity:

Sound level or **intensity** I is the amount of power per square metre (in W m⁻²). This is a *linear* measure of sound intensity. Doubling the Watts per square metre doubles the sound intensity. However, the ear can detect sounds which vary in intensity over a wide range. If we were to use W m⁻² as a measure of sound intensity experienced by the ear, then the numbers would go from very small (1×10^{-12} W m⁻²) to very large (1×10^{2} W m⁻²). This is very inconvenient, so we usually express sound intensity using **decibels (db)** which is a logarithmic scale.

For a sound intensity I in W m⁻², the sound intensity in db is given by:

$$\boxed{\beta = 10\log_{10}\frac{I}{I_o}}$$

Sound	Typical level
whisper	20 db
ordinary conversation	65 db
motor car	60 db
train	90 db

$I_o = 1 \times 10^{-12}$ W m⁻² and is approximately the threshold of hearing at 1kHz

2.3.12 Hearing

The human ear is subjective when it comes to interpreting pressure
variations as sound. Objective quantities such as **intensity**, and **frequency**
are measurable using scientific instruments. Subjective qualities such as
loudness and **pitch** are not measurable using scientific instruments.

Objective	Subjective
Intensity • measured in decibels and is an indication of the energy carried by the sound wave. The intensity depends on the amplitude A of the sound waves.	**Loudness** (volume) • interpreted as intensity by the ear but is frequency dependent. For a given intensity, sounds at low frequency or very high frequency are not as loud as those at moderate frequencies (a few kHz).
Frequency • measured in cycles per second.	**Pitch** • interpreted as frequency by the ear but depends on the intensity. The pitch of a single frequency, or a "pure tone" becomes lower as the intensity increases.
Timbre • harmonic content	**Tonal quality** • depends on the component frequencies of the sound. A pure tone has a "single" frequency. Bright sounds have more power in the high frequency components (to musicians, sharp sounds will mean higher pitched).

2.3.13 Reflection and absorption of sound

When energy is lost from the sound wave as it passes through the medium we say that the sound has undergone **absorption**. The absorption of sound in most materials increases as with increasing frequency of the sound waves and also increases with decreasing density of the material.

What happens when a loud noise is produced inside a room?

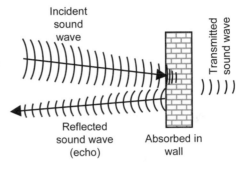

Incident sound wave

Transmitted sound wave

Reflected sound wave (echo)

Absorbed in wall

• Sound waves strike the wall.

• Some of the sound is reflected by the wall and it is called an echo.

• Whatever doesn't get reflected goes through into the wall and undergoes some degree of absorption.

During absorption, the energy of the sound wave is converted to heat within the medium.

• Anything that is not completely absorbed, gets transmitted through to the other side of the wall.

For large open spaces, absorption within the medium is important. For enclosed spaces, reflection and absorption at the walls is important.

How do we know what is reflected and what goes on through the material? When sound goes from one medium to another, there is a large reflection component when there is a large difference in **acoustic impedance** of the two materials. The acoustic impedance is given by: $z = \rho c$

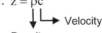

Velocity

Density

For sound going from air and striking a brick wall, there is a very large acoustic impedance mismatch and most of the sound is reflected. For sound waves striking a wall covered with a cloth, there is less of an echo because there is less of an acoustic impedance mismatch and the sound is absorbed by the cloth. So with a brick wall, sound is easily reflected, but any that does go into it is readily absorbed!

Material	z kgm⁻¹s⁻¹
Air	415
Water	1.5×10^6
Concrete	8×10^6
Steel	33×10^6

2.3.14 Intensity variations

1. Plane waves

No divergence of rays.
Wavefronts are parallel
planes. Same energy
passes through same
are, thus no variation in
intensity.

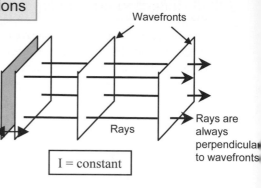

Wavefronts

Rays

Rays are
always
perpendicular
to wavefronts

$$I = \text{constant}$$

2. Cylindrical waves

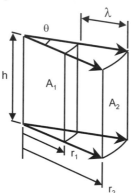

$A_1 = r_1 \theta h$

$A_2 = r_2 \theta h$

$I_1 = \dfrac{P}{A_1}$

$I_2 = \dfrac{P}{A_2}$

$A = f(r)$

Expanding wavefront
increases surface area
with distance from
source:

$$I \propto \frac{1}{r}$$

3. Spherical waves

Area of sphere = $4\pi r^2$

If the distance r
from the spherical
wavefront to the
source is doubled,
then the area
increases by 4 times.
Hence:

Point
source

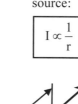

A_1

A_2

$$I \propto \frac{1}{r^2}$$

function of r^2

The direction in which waves radiate from the source is shown by straight lines called
rays. A **wavefront** is a surface on which all points have the same phase of
oscillation and is normal to the rays.

2.3.15 Examples

1. Calculate the amplitude of the superposition of the two waves which are described by the equations below at a position x = 2 m:

$$y_1 = 20\sin(8t - 5x)$$
$$y_2 = 20\sin(8t + 5x)$$

Solution:

$$5 = \frac{2\pi}{\lambda}$$

$$\lambda = 1.256\,m$$

$$y_{1+2} = 2(20)\cos\frac{2\pi 2}{1.256}$$

$$= -33.45\,m$$

2. For the superposition of the waves in the previous question, determine the position x of the first antinode (assuming x = 0 is a node).

Solution:

$$5 = \frac{2\pi}{\lambda} = 1.256\,m$$

$$x = \frac{1.256}{2}$$

$$= 0.628\,m$$

3. Calculate the fundamental frequency of an air filled tube which is open at both ends, has a diameter of 10 mm, and a length of 400 mm and is at 0 °C.

Solution:

$$L = \frac{\lambda}{2} \qquad v = \sqrt{\frac{\gamma RT}{M}}$$

$$\lambda = 2(0.4) \qquad = \sqrt{\frac{(1.4)(8.314)(273)}{0.028}}$$

$$= 800mm$$

$$= 336\,ms^{-1}$$

$$v = f\lambda$$

$$f = \frac{336}{0.8}$$

$$= 421Hz$$

2.4 Light

Summary

$$\frac{\sin \theta_i}{\sin \theta_r} = \frac{n_r}{n_i}$$ Snell's law

$$\sin \theta_C = \frac{n_r}{n_i}$$ Critical angle

$$n = 1 + \frac{Nq_e^2}{2\varepsilon_o m\left(\omega_o^2 - \omega^2\right)}$$ Dispersion equation

2.4.1 Light rays

Consider a light source:

1. The directions in which light waves radiate from the source are shown by straight lines called **rays**.

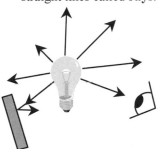

2. Rays of light emanate from the source and radiate outwards in all directions.

3. Some rays enter an observer's eye and are focussed by the lens in the eye to form an image on the rctina.

4. Other rays are reflected and/or absorbed by the surroundings

Rays are imaginary lines drawn perpendicular to the wavefronts and indicate the direction of travel of the waves

In this example, the light source can be considered a **point source** emitting spherical waves. Light rays thus radiate outwards. Far away from the source, the radius of the wavefronts are very large and can be approximated by **plane waves** in which the rays are parallel.

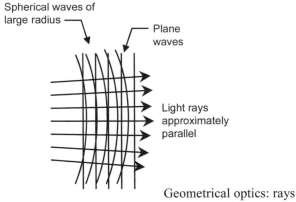

Spherical waves of large radius

Plane waves

Light rays approximately parallel

Geometrical optics: rays
Physical optics: waves

2.4.2 Reflection

If the surface is smooth, the angle of
reflection = the angle of incidence.

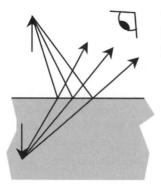

The eye sees the light as
if it were coming from a
point behind the surface.

This type of reflection is
called **specular reflection**.

If the surface is rough, the angle of reflection still equals the angle of
incidence for each ray, but these angles are now such that the light is
scattered.

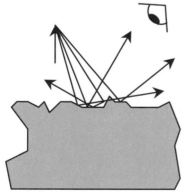

Many of the light rays do not
enter the eye at all and an image
cannot be formed. Or, the ones
that do enter the eye appear not
to come from a single point so
again, no image is formed.

This type of reflection is called
diffuse reflection.

2.4.3 Refraction

If a light wave strikes a medium (such as a block of glass) at an angle, then the waves which enter the medium slow down and the wavelength becomes shorter to compensate. This has the effect of altering the direction of travel of the wave.

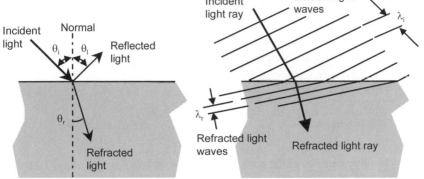

The speed of light in a material is always less than that of the speed of light in a vacuum. The ratio of these two speeds is called the **refractive index**.

$$n = \frac{c}{v}$$

The frequency of light does not change when the wave passes from one material to another. The velocity does change, hence, since $v = f\lambda$, the wavelength also changes.

Experiments show that the angle of incidence = angle of reflection

$$\theta_i = \theta_l$$

→ All angles measured w.r.t. the normal to the surface

Experiments show that for monochromatic light, the angles of incidence and refraction are related by the refractive indices of the two materials:

$$n_i \sin \theta_i = n_r \sin \theta_r$$

$$\boxed{\frac{\sin \theta_i}{\sin \theta_r} = \frac{n_r}{n_i}} \quad \textbf{Snell's law}$$

Medium	n
Vacuum	1
Air at STP	1.0003
Glass	1.52

Optically less dense ↓

Optically more dense ↓

If light travels from a material with low value of n to one with high value of n, then the velocity of the wave is reduced and the wavelength becomes shorter to compensate. The frequency remains the same. The path taken by light rays is reversible. It doesn't matter whether the light rays travel from an optically more dense to less dense material or vice versa.

2.4.4 Total internal reflection

Let us examine the path of light rays when passing from an optically more dense to a less dense medium ($n_i > n_r$).

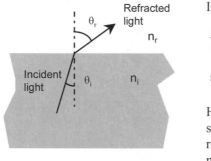

In keeping with Snell's law:

$$\frac{\sin \theta_i}{\sin \theta_r} = \frac{n_r}{n_i}$$

$$\sin \theta_r = \frac{n_i}{n_r} \sin \theta_i$$

Here we have $n_i/n_r > 1$ hence $\sin \theta_r > \sin \theta_i$ and thus $\theta_r > \theta_i$ and the refracted ray is bent away from the normal.

What happens when the angle of incidence is made increasingly larger?

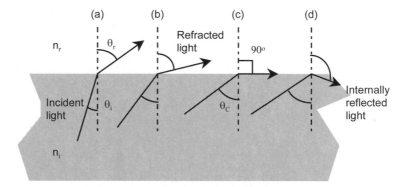

As θ_i is made larger, there comes a point where the angle of refraction $\theta_r = 90°$ and the refracted ray grazes the surface of the boundary between the two mediums. This angle of incidence is called the **critical angle**. Thus:

$$\sin \theta_C = \frac{n_r}{n_i}$$

At angles of incidence greater than the critical angle, the refraction no longer takes place (since by **Snell's law**, $\sin \theta_r > 1$ which is not possible). The ray is then totally internally reflected.

2.4.5 Dispersion

The **speed of light** in a vacuum is the same for all wavelengths. The speed of light in a medium depends on the wavelength as well as the properties (the permittivity) of the medium. ⟶ Because of the interaction between electrons in the medium and the action of the electric field of the light waves.

Since the refractive index is a measure of the relative speeds of light waves in a vacuum and a medium, then the value of n for a medium thus depends on the wavelength or the frequency of the incoming light.

Consider white light incident on a transparent medium. When light is incident on the medium, the incoming electric field E causes a distortion of the internal charge distribution of the molecules (polarisation). This causes molecules to try to align themselves with the E field – thus altering the net field within the material. For a rapidly varying

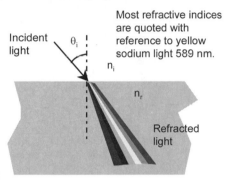

Most refractive indices are quoted with reference to yellow sodium light 589 nm.

E field, the molecules may not be able to move fast enough to keep up with the changing field. The multitude of frequencies in the incoming white light above results in different responses, on an atomic scale, within the material. The result is that for higher frequency components, the velocity of the light is reduced and the refractive index is increased. This effect can be quantified by the **dispersion equation**.

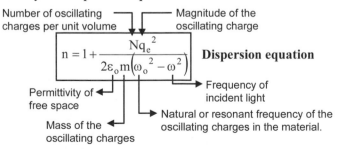

Number of oscillating charges per unit volume

Magnitude of the oscillating charge

$$n = 1 + \frac{Nq_e^2}{2\varepsilon_0 m\left(\omega_o^2 - \omega^2\right)}$$ **Dispersion equation**

Permittivity of free space

Mass of the oscillating charges

Frequency of incident light

Natural or resonant frequency of the oscillating charges in the material.

The **dispersion equation** gives n for variations in frequency of incident light. This frequency itself does not change for the refracted wave, but wavelength and velocity do. If everything except ω is held constant, then for an increase in ω, (decrease in λ) n also increases.

2.4.6 Example

1. A light at the bottom of a 2m deep swimming pool is switched on.
    What is the diameter of the beam at the surface of the water if
    $n_{water}=1.33$?

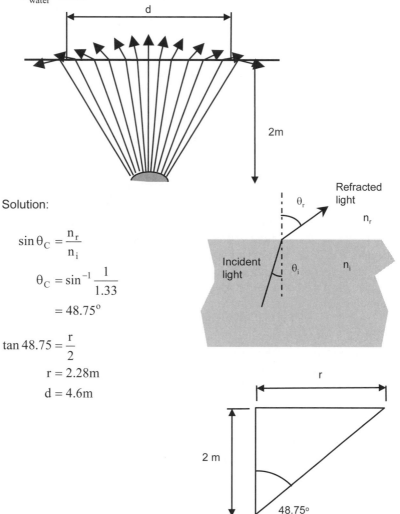

Solution:

$$\sin \theta_C = \frac{n_r}{n_i}$$

$$\theta_C = \sin^{-1} \frac{1}{1.33}$$

$$= 48.75^{\circ}$$

$$\tan 48.75 = \frac{r}{2}$$

$$r = 2.28m$$

$$d = 4.6m$$

2.5 Mirrors

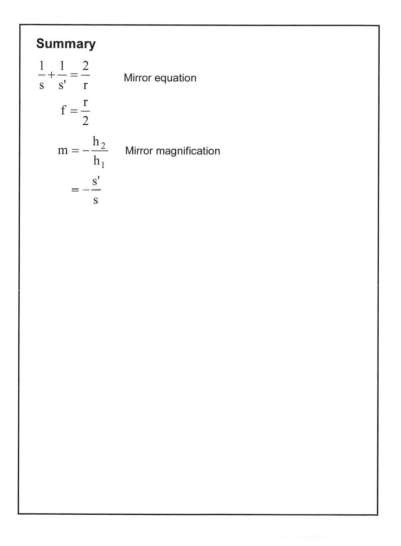

Summary

$$\frac{1}{s} + \frac{1}{s'} = \frac{2}{r}$$ Mirror equation

$$f = \frac{r}{2}$$

$$m = -\frac{h_2}{h_1}$$ Mirror magnification

$$= -\frac{s'}{s}$$

2.5.1 Mirrors

Mirrors have two sides:

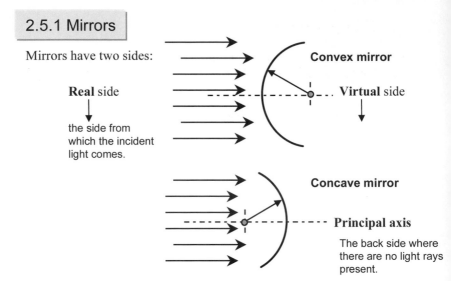

Real side

the side from
which the incident
light comes.

Convex mirror

Virtual side

Concave mirror

Principal axis

The back side where
there are no light rays
present.

Light rays coming from objects are reflected in mirrors to form an image
What we see in a mirror:

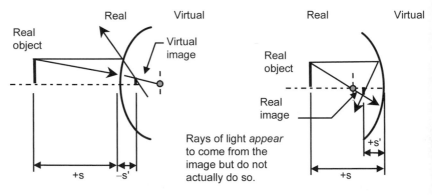

The **object distance** "s" is positive and the object is real if the object is on
the real side. If the object appears to exist on the virtual side, the object
distance is negative and the object is a **virtual object**.

A **real object** is one where the rays of light actually come from the object.
A virtual object is one where rays of light *appear* to come from the object
but do not actually do so.

The **image distance** "s' " is positive and the image is "real" if the image is
on the real side. The image distance is negative and virtual if it lies on the
virtual side.

2.5.2 Sign conventions

Radius of curvature: The radius of curvature "r" is positive if the centre of curvature "c" is on the real side, and negative if the centre of curvature is on the virtual side.

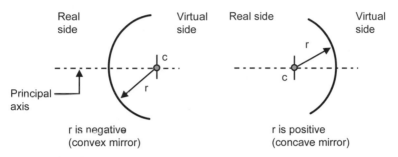

r is negative
(convex mirror)

r is positive
(concave mirror)

Focal length: Any incident rays which are parallel to the principal axis are reflected such that the reflected rays appear to come from a point half way between the surface of the mirror and the centre of curvature (this is for a spherical mirror only - not a lens).

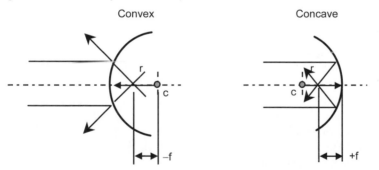

Conversely, any rays passing through, or travelling towards, the focal point will be reflected parallel to the principal axis.

The focal length (f) is positive if the centre of curvature lies on the real side and negative if the centre of curvature lies on the virtual side. The focal length f has the same sign as r.

Convex mirror: f is negative
Concave mirror: f is positive

2.5.3 Ray diagrams

The path of two light rays through a mirror system are known no matter what kind of mirror or object is involved. The two rays are:

1. Any incident rays which are parallel to the principal axis are reflected such that the reflected rays appear to come from the **focal point** "f".

2. All rays passing through (or travelling towards) the centre of curvature "c" of the mirror strike the mirror at right angles and hence are reflected back along their original path.

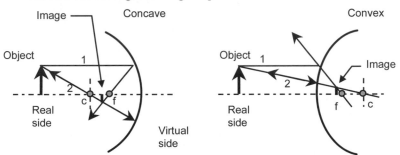

The eye interprets the light rays emanating from a point in space as coming from an **object**. The image formed by the mirror is the object for the lens of the eye.

The eye "sees" the image of the actual object in the mirror. The image may be magnified or diminished in size, inverted or upright in orientation, depending on where the object is situated with respect to the mirror and the curvature of the mirror.

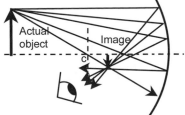

The object distance, image distance, focal length and radius of curvature are related by:

$$\frac{1}{s} + \frac{1}{s'} = \frac{2}{r}$$

$$f = \frac{r}{2}$$

This formula only applies to rays which make a small angle to the principal axis (paraxial rays).

The **linear magnification** "m" is the ratio of the image size to the object size:

$$m = -\frac{h_2}{h_1}$$

$$= -\frac{s'}{s}$$

If m is +ve, then image is upright if m is -ve, then image is inverted

2.5.4 Example

1. Professor Smith uses a concave shaving mirror with a focal length of 450 mm. How far away should Prof. Smith's face be from the mirror for him to see an image of his face which is upright, and twice its actual size?

Solution:

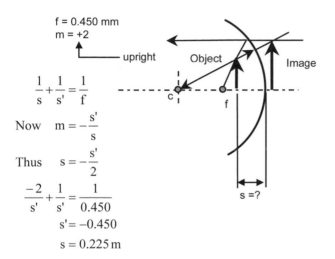

$f = 0.450$ mm
$m = +2$

$$\frac{1}{s} + \frac{1}{s'} = \frac{1}{f}$$

Now $m = -\dfrac{s'}{s}$

Thus $s = -\dfrac{s'}{2}$

$$\frac{-2}{s'} + \frac{1}{s'} = \frac{1}{0.450}$$

$$s' = -0.450$$

$$s = 0.225\,\text{m}$$

2.6 Lenses

Summary

$$\frac{1}{s} + \frac{1}{s'} = \frac{1}{f}$$ Thin lens equation

$$\frac{1}{f} = \left[\frac{n_2}{n_1} - 1\right]\left[\frac{1}{r_1} - \frac{1}{r_2}\right]$$ Lens maker's equation

$$m = -\frac{h'}{h} = -\frac{s'}{s}$$ Linear magnification

$$P = \frac{1}{f}$$ Lens power

$$M = \frac{s_o}{s}\left(1 + \frac{1}{f}(s-1)\right)$$ Angular magnification

$$M = \frac{s_o}{f}$$ Angular magnification

$$M = \frac{s_o}{f} + 1$$ Angular magnification

2.6.1 Thin lens

Rays passing through a transparent material with parallel sides are displaced, but not deviated in direction.

Rays passing through a transparent material with nonparallel sides are deviated in direction.

The surfaces of a lens are shaped so that all parallel rays are deviated so that they meet at a single point.

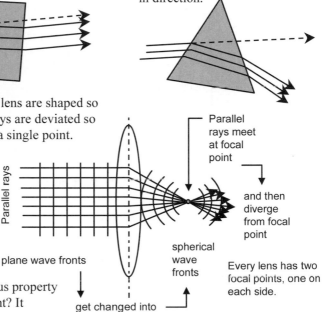

Parallel rays meet at focal point

and then diverge from focal point

Every lens has two focal points, one on each side.

Parallel rays

plane wave fronts

get changed into

spherical wave fronts

Why is this curious property of a lens important? It underlies the operation of:

- optical instruments
- eyes
- eye glasses and contact lenses

A **thin lens** is one whose spherical surfaces have a radius large in comparison with the thickness of the lens. Equations for thin lenses are relatively simple.

2.6.2 Lens action

Light rays travel in straight lines. When an object is illuminated, or is
self-luminous (e.g. a candle or light bulb), rays of light emanate from the
object.

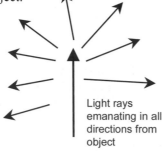

What happens to these light rays?
Some of them go off and strike other
surfaces and are reflected, refracted
or absorbed. Some go off into space,
and some may enter someone's eye.

Light rays
emanating in all
directions from
object

Now, if it is desired to form an image of the object on a screen (say a
projector screen, or the retina of the eye), then a lens is necessary. Consider
the rays of light which emanate from the top of the arrow shown below:

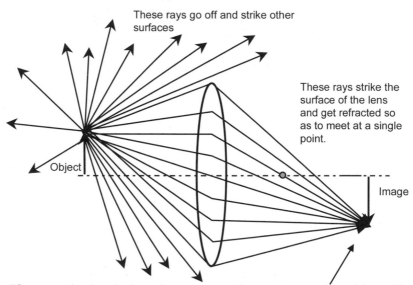

These rays go off and strike other
surfaces

These rays strike the
surface of the lens
and get refracted so
as to meet at a single
point.

Object

Image

If a screen is placed where the rays meet, then an image of the object will
appear on the screen. Only those rays which are intercepted by the lens
contribute to forming the image. A larger lens gathers more of the rays and
so the image is brighter. Large telescopes have large lenses so as to gather
as many of the light rays coming from a faint star as possible.

2.6.3 Lenses

A lens consists of two refracting surfaces. Here we consider two spherical surfaces which are positioned close together - a **thin lens**.

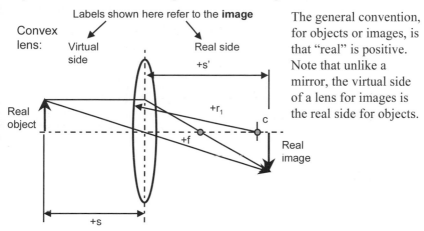

The general convention, for objects or images, is that "real" is positive. Note that unlike a mirror, the virtual side of a lens for images is the real side for objects.

For light rays travelling from left to right through a lens, an image is "real" if it is formed on the right hand side of the lens and "virtual" if formed on the left hand side. These terms are just labels for identifying which side of the lens the image or objects are. If the rays diverge from an object, the object is real and is on the left side of the lens. If the rays (going from left to right) converge to the object, then the object is virtual and lies on the right side of the lens.

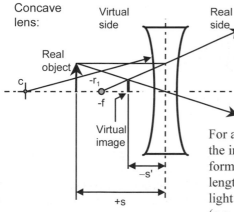

The image shown here is a **virtual image** since it is formed on the left hand side of the lens and thus the image distance s' is negative.

The radius of curvature is +ve if it lies on the real side. In all formulas, r_1 is the radius of curvature of the first surface upon which light is incident.

For a lens, the **focal length** is positive if the incident parallel light converges to form a real image (e.g. convex). Focal length is negative if incident parallel light diverges to form a virtual image (e.g. concave).

2.6.4 Ray diagrams

The path of two light rays through a **thin lens** system are known no matter what kind of lens or object is involved. The two rays are:

1. A ray that passes through the centre of a lens continues on in a straight line.

2. A ray that travels parallel to the principal axis will emerge from the lens in a direction towards the focal point (and vice versa).

Note, compared to a mirror: a lens has two focal points, one on each side of the lens and two radii of curvature. r_1 is the radius of curvature of the first surface upon which light is incident. The focal length is *not* one half of the radius of curvature for a lens. The two focal lengths for a lens are equal, even if the radii of curvature for each side are different (if the medium is the same on each side of the lens e.g. air).

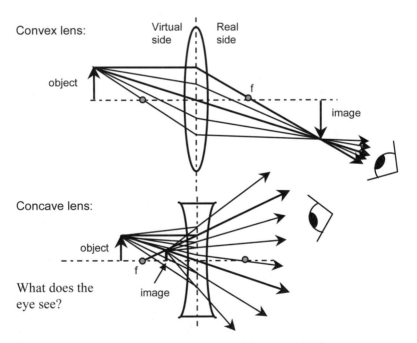

Convex lens:

Virtual side

Real side

object

f

image

Concave lens:

object

f

image

What does the eye see?

The eye interprets the light rays emanating from a point in space as coming from an "object". The image formed by the lens is the object for the lens of the eye. Compared to the actual object, the image may be magnified or diminished in size, inverted or upright in orientation, depending on where the object is situated with respect to the lens and the nature of the lens.

2.6.5 The eye

1. Object is at a distance s from the lens of the eye.

Note that most of the focussing is really done by the **cornea**. Lens provides only a "fine" adjustment.

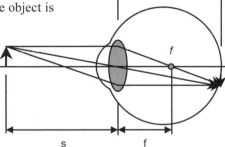

The **retina** of the eye is a "screen" upon which a real image is focussed. Image distance s' is fixed by the dimensions of the eye.

2. Examine what happens if the object is brought closer to the eye:

Eye muscles for lens encircle the lens. Contraction causes lens to bulge more, r and f decrease. When muscle relaxed, f is largest.

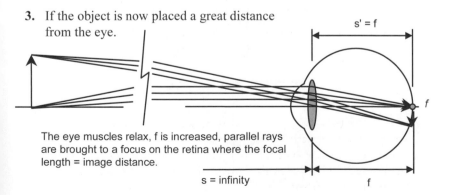

For image to be in focussed on the retina, focal length is made smaller (eye muscles contract and increase curvature of lens) so that image distance s' remains fixed (see thin lens eqn later)

3. If the object is now placed a great distance from the eye.

The eye muscles relax, f is increased, parallel rays are brought to a focus on the retina where the focal length = image distance.

s = infinity

2.6.6 Near point and far point

The **near point** is the closest distance that an object may be placed in front of the eye which may be brought into focus on the retina. Usually about 250mm for a normal eye. This distance is limited by the elasticity of the crystalline lens in the eye.

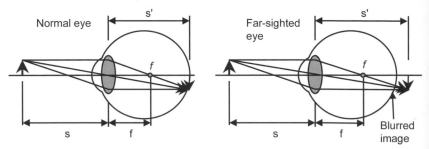

For **far-sighted** people, the near point is much greater than 250 mm and objects far away from the eye can only be focussed. Light from nearby objects is focussed beyond the retina. A correcting lens (convex) is required to allow nearby objects to be brought into focus.

The **far point** is the furthest distance that an object may be placed for its image to be focussed on the retina (usually infinity).

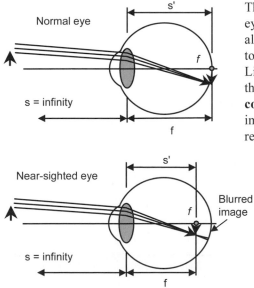

The lens in a **near-sighted** eye cannot relax enough to allow parallel incoming light to be focussed on the retina. Light is focussed in front of the retina. A diverging or **concave lens** is required for image to be focussed on retina.

2.6.7 Lens equations

The object distance, image distance, focal length and radius of curvature
are related by the thin lens and lens-makers equations:

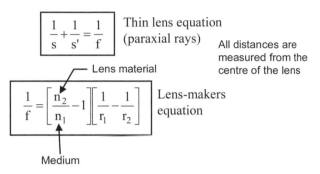

$$\frac{1}{s}+\frac{1}{s'}=\frac{1}{f}$$

Thin lens equation
(paraxial rays)

All distances are
measured from the
centre of the lens

— Lens material

$$\frac{1}{f}=\left[\frac{n_2}{n_1}-1\right]\left[\frac{1}{r_1}-\frac{1}{r_2}\right]$$

Lens-makers
equation

↑
Medium

The **linear magnification** "m" is the ratio of the image size to the
object size:

$$m=-\frac{h'}{h}$$
$$=-\frac{s'}{s}$$

If m is positive, then
image is upright.
if m is negative,
then image is
inverted.

The power of a thin lens in air is given by the reciprocal of the focal length
and has units **dioptres**.

$$P=\frac{1}{f}$$

└— metres

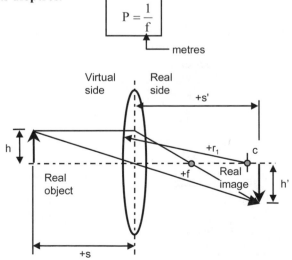

2.6.8 Magnifying glass

1. If the object is placed a *little* further away from the focal
    point, then the image is real, inverted and magnified.

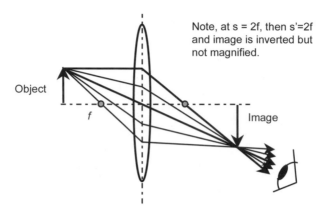

Note, at s = 2f, then s'=2f
and image is inverted but
not magnified.

Object

f

Image

2. If the object to be examined is positioned a little closer than the focal
    point. This causes the rays to diverge from the lens thus forming a
    virtual image behind the actual object.

3. The eyes interpret light rays as if they are emanating from a larger
    object at distance s'.

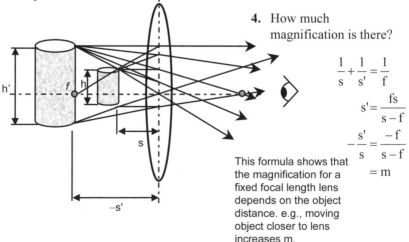

4. How much
    magnification is there?

$$\frac{1}{s}+\frac{1}{s'}=\frac{1}{f}$$

$$s'=\frac{fs}{s-f}$$

$$-\frac{s'}{s}=\frac{-f}{s-f}$$

$$=m$$

This formula shows that
the magnification for a
fixed focal length lens
depends on the object
distance. e.g., moving
object closer to lens
increases m.

2.6.9 Magnification

The **magnifying power** of an optical instrument is defined as the **angular magnification**:

$$M = \frac{\theta_i}{\theta_o}$$

Objects can be made to appear larger by bringing them closer to the eye. But the eye can only accommodate objects bought to a distance not less than the **near point** (usually about 250 mm) .

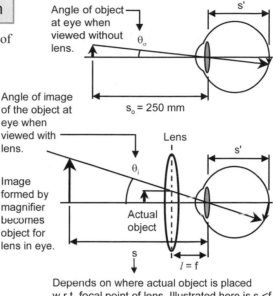

Angle of object at eye when viewed without lens.

$s_o = 250$ mm

Angle of image of the object at eye when viewed with lens.

Image formed by magnifier becomes object for lens in eye.

Lens

Actual object

$l = f$

Depends on where actual object is placed w.r.t. focal point of lens. Illustrated here is s <f so that virtual upright image is obtained.

In an optical instrument, the important thing is the size of the image formed on the retina. This determines the apparent size of the object. It can be shown that:

$$M = \frac{s_o}{s}\left(1 + \frac{1}{f}(s - l)\right)$$

≈ 250mm

where s_o is the distance from the object to the eye (usually taken to be the near point) without lens, and f is the focal length of the lens. When the lens is placed so that the eye is at the focal point $l = f$, or the object is placed at the focal s = f point, then:

$$M = \frac{s_o}{f}$$

Magnifying power can be increased by decreasing the focal length. But, various aberrations limit the value of M to about 3X to 4X for a single convex lens. If more magnifying power is needed, then need to use a compound arrangement of lenses.

With the magnifying lens in place (here shown at $l = f$), the object may be bought closer to the eye to increase M.

If eye is bought up close to the lens ($l = 0$) and image at $s = s_o$, then:

$$M = \frac{s_o}{f} + 1$$

For a simple magnifier, what do you see when the object is placed *at* the focal point of the lens?

Consider the thin lens equation with s = f.

$$\frac{1}{f} + \frac{1}{s'} = \frac{1}{f}$$

$$\frac{1}{s'} = 0$$

s' = ∞ the image is formed at infinity!

What about the linear magnification?

$$m = -\frac{s'}{s}$$

= ∞ the image is infinitely large!

"At infinity" means that the object (for the eye) is sufficiently far away so that light rays from a point on the object arrive at the eye are virtually parallel.

The image becomes an "object" for the lens of the eye. The object (as far as the eye is concerned) is now at a very large distance from the eye. But, no matter how far away, it still subtends some (small) angle.

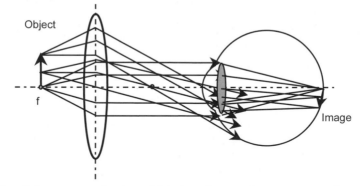

Parallel rays enter the eye which then focuses these rays onto the retina. The focal length of the lens in the eye is thus the diameter of the eyeball in this instance and the eye muscle is relaxed - most comfortable viewing.

Remember, it is the **angular magnification** which is perceived by the eye. For an image at infinity, the angular magnification depends only on f (i.e. with reference to a standard near point of 250 mm).

$$M = \frac{s_o}{f}$$

2.6.10 Lens aberrations

Chromatic aberration

We have seen that the refractive index of materials depends upon the wavelength of the light being refracted since the speed of light in a medium depends on the wavelength as well as the properties of the medium.

Thus, when incident white light is refracted by the lens surfaces, small wavelengths (e.g. blue) will be refracted at greater angles at each surface than long wavelengths (red light).

The refractive index of the lens material depends on the wavelength of the light being refracted. For optical instruments, best to use yellow light to calculate focal lengths and radii of curvature.

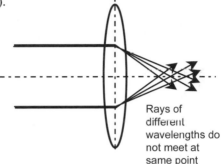

Rays of different wavelengths do not meet at same point

Spherical aberration

Formulas and discussions so far have assumed that rays of light passing through the *spherical* lens are paraxial (i.e. make small angles with the axis of the lens). This is not always the case as shown below:

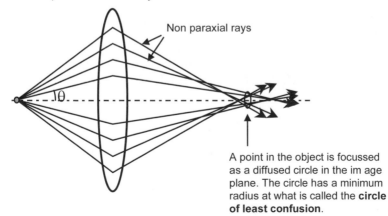

Non paraxial rays

A point in the object is focussed as a diffused circle in the im age plane. The circle has a minimum radius at what is called the **circle of least confusion**.

This is because in the derivation of the thin lens equation, it is assumed that $\sin\theta = \theta$ in **Snell's law** so that $n_1\theta_1 = n_2\theta_2$.

2.6.11 Example

1. George has trouble seeing the blackboard in Physics lectures. He gets his eyes tested and finds that he can only see objects clearly if they are less than 100 mm away from him.

    (a) What power of lens would George need for spectacles to enable him to see objects that are far away?

    (b) If glass of refractive index 1.55 is to be used to make the lenses, what radius of curvature is required (assume $R_2 = -R_1$)?

Solution:

(a) $s' = -0.1\,m$

    $s = \infty$

    $\dfrac{1}{f} = \dfrac{1}{-0.1} + 0$

    $f = -0.1\,m$

    $= -10$ Dioptres

    ↑

    Concave

(b) $\dfrac{1}{f} = (n-1)\left(\dfrac{1}{R_1} - \dfrac{1}{R_2}\right)$

    $\dfrac{1}{-100} = (1.55 - 1)\left(\dfrac{1}{R_1} - \dfrac{1}{-R_1}\right)$

    $\dfrac{2}{R} = \dfrac{1}{100(0.55)}$

    $R_1 = -110\,mm$

    $R_2 = 110\,mm$

Note, objects at infinity need to appear as if they were 0.1 m from George's eyes for him to see them sharply. s' is negative since image formed by spectacles will be on virtual side of lens and will serve as an object for the lens in the eye.

2.7 Optical instruments

Summary

$$M_{Total} = m_e M_o \qquad \text{Microscope}$$

$$= \frac{250 s_o'}{f_e f_o}$$

$$M = \frac{f_o}{f_e} \qquad \begin{array}{l}\text{Astronomical}\\ \text{telescope}\end{array}$$

2.7.1 Optical instruments

In an optical instrument, the eye is usually bought very close to the **eyepiece** or **ocular**. In this case, the **angular magnification** is:

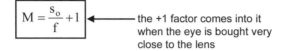

$$M = \frac{s_o}{f} + 1$$ the +1 factor comes into it when the eye is bought very close to the lens

For a **single lens magnifier**, this is usually limited to about 4X due to lens aberrations. How then to increase the magnification of an object and retain a good quality image? We use lenses in combination. There are two ways of combining lenses:

1. Thin lenses in contact

When two thin lenses are placed very close together, we say they are "in contact". The power of the combination of the lenses is the sum of the powers of each lens.

$$\frac{1}{f} = \frac{1}{f_1} + \frac{1}{f_2}$$ This formula applies to combinations of +ve (convex) and -ve (concave) lenses.

2. Thin lenses not in contact

Combinations of lenses not in contact enables optical instruments of varying design to be constructed.

The general procedure is to treat the *image made by the first lens as the object for the second lens*. One must be very careful about image and object distances, since some objects must be classed as being "virtual" or "real" depending on which side of the lens they exist.

2.7.2 Microscope

Each lens is treated separately. Rays are *not* drawn through lens 1 and then through lens 2 in one operation. Rays can only be drawn using the two known directions of rays (through focus and centre) in a step-by-step procedure.

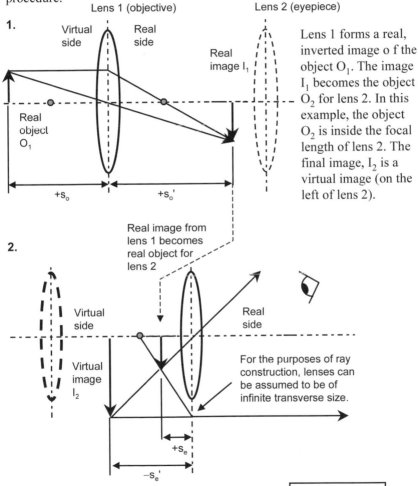

1. Lens 1 (objective)   Lens 2 (eyepiece)

Virtual side — Real side

Real image I_1

Real object O_1

$+s_o$ $+s_o'$

Lens 1 forms a real, inverted image o f the object O_1. The image I_1 becomes the object O_2 for lens 2. In this example, the object O_2 is inside the focal length of lens 2. The final image, I_2 is a virtual image (on the left of lens 2).

2.

Real image from lens 1 becomes real object for lens 2

Virtual side

Virtual image I_2

Real side

For the purposes of ray construction, lenses can be assumed to be of infinite transverse size.

$+s_e$

$-s_e'$

For the largest magnification (i.e. retinal image), the final image should be placed at the near point ($s_e' = 250$mm). It can be shown that the total angular magnification is:

$$M_{Total} = m_e M_o$$
$$= \frac{250 s_o'}{f_e f_o}$$

2.7.3 Astronomical telescope

In an astronomical telescope, the object O_1 is usually at infinity hence the image I_1 is formed at the focal length of the **objective** lens (i.e. lens 1). For relaxed viewing, the final image should also be at infinity, hence, lens 2 (the **eyepiece**) is placed so that the object O_2 is at f_2 ($= f_e$).

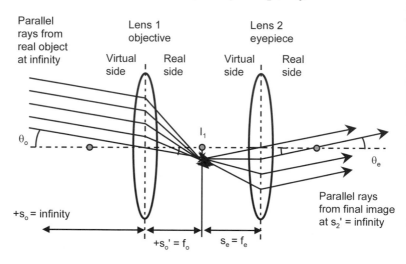

If h' is the height of the image I_1, then:

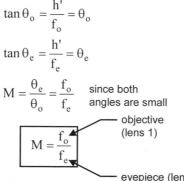

$$\tan\theta_o = \frac{h'}{f_o} = \theta_o$$

$$\tan\theta_e = \frac{h'}{f_e} = \theta_e$$

$$M = \frac{\theta_e}{\theta_o} = \frac{f_o}{f_e} \quad \text{since both angles are small}$$

objective (lens 1)

$$M = \frac{f_o}{f_e}$$

eyepiece (lens 2)

We might well ask, if the final image is at infinity, how then do we see anything? We will get an image on a screen if the screen is placed at infinity. But, if the rays enter our eye, then the lens in the eye focuses the rays on to the screen of our retina.

The **angular magnification** is the ratio of the angle subtended by the final image I_2 formed by lens 2 at the eye θ_e to the angle θ_o subtended by the object O_1 at the objective lens (lens 1) (which is the same as that subtended by the object at the unaided eye).

2.7.4 Special case

A particularly interesting case arises when the image formed by the first lens (the objective) is not between the two lenses in combination. For example, with the lens combination shown below, with lens 2 not in position, the image I_1 is to the right of the position of lens 2. I_1 then becomes a virtual object for lens 2.

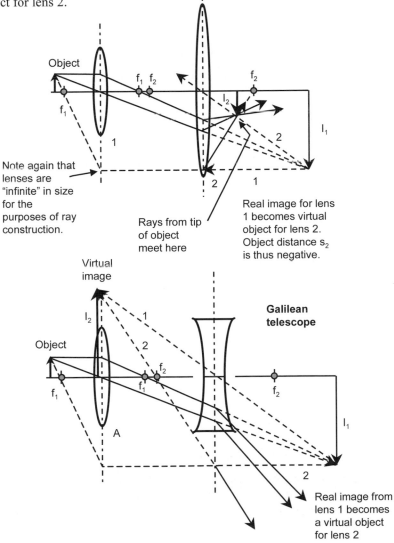

Note again that lenses are "infinite" in size for the purposes of ray construction.

Rays from tip of object meet here

Real image for lens 1 becomes virtual object for lens 2. Object distance s_2 is thus negative.

Virtual image

Object

Galilean telescope

Real image from lens 1 becomes a virtual object for lens 2

2.7.5 Example

2. A convex lens (f = 300 mm) is placed 200 mm from a concave lens
   (f = −50mm). An object is placed 6 m away from the convex lens.
   Determine the position, magnification and nature of the final image.

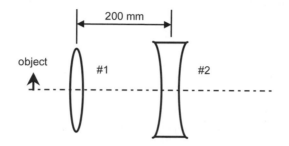

Solution:

$$\frac{1}{300} = \frac{1}{6000} + \frac{1}{s'}$$

$$s' = 315.8\,mm$$

$$s_2 = 200 - 315.8$$

$$= -115$$

$$-\frac{1}{50} = \frac{-1}{115} + \frac{1}{s'}$$

$$s' = -88\,mm$$

$$m_1 = -\frac{315}{6000}$$

$$= -0.053$$

$$m_2 = -\frac{-88}{-115}$$

$$= -0.756$$

$$m_T = m_1 m_2 = 0.041$$

88 mm virtual side of lens 2 with magnification 0.041

2.8 Interference

Summary

$$d \sin \theta = \frac{2n+1}{2} \lambda$$ Destructive interference

$$n = 0,1,2,...$$

$$d \sin \theta = n\lambda$$ Constructive interference

$$n = 0,1,2,...$$

$$I = I_{max} \cos^2 \left(\frac{\pi}{\lambda} d \sin \theta \right)$$ Fringe intensity

$$\Delta x = m\lambda$$

$$= 2d$$ Thin film interference maxima

$$\Delta x = \left(m + \frac{1}{2} \right) \lambda$$ Thin film interference minima

$$= 2d$$

2.8.1 Interference

Interference arises from the **superposition** of waves.

Constructive	**Destructive**	— light of a "single" wavelength
Amplitude of resultant is greater than components.	Amplitude of resultant is less than components.	┌ phase difference (if any) between them remains constant.

Consider two point sources S_1 and S_2 of monochromatic, coherent light separated by a distance d emitting spherical wave fronts.

Ray 1 has a greater path to travel to the screen compared to ray 2. The **path difference** is Δx.

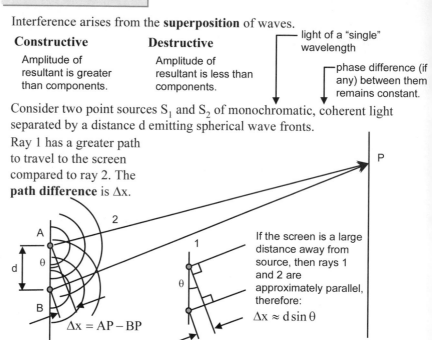

If the screen is a large distance away from source, then rays 1 and 2 are approximately parallel, therefore:

$$\Delta x \approx d \sin \theta$$

$$\Delta x = AP - BP$$

Because of this path difference, the rays, which may be initially in phase, may arrive at P out of phase. If the **phase difference** $\Delta\phi$ at P is 180° (π), then destructive interference occurs. The distance Δx must thus be equal to $\lambda/2$. If the phase difference at P is 0, then constructive interference and the distance $\Delta x = \lambda$.

$$y_1 = A \sin(\omega t - kx)$$
$$y_2 = A \sin(\omega t - kx + \Delta\phi)$$

The phase difference arises from difference in path length $\Delta x = d\sin\theta$:

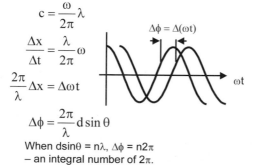

$$c = \frac{\omega}{2\pi}\lambda$$

$$\frac{\Delta x}{\Delta t} = \frac{\lambda}{2\pi}\omega$$

$$\frac{2\pi}{\lambda}\Delta x = \Delta\omega t$$

$$\Delta\phi = \frac{2\pi}{\lambda}d\sin\theta$$

$\Delta\phi = \Delta(\omega t)$

When $d\sin\theta = n\lambda$, $\Delta\phi = n2\pi$ – an integral number of 2π.

Destructive interference

$$d\sin\theta = \frac{2n+1}{2}\lambda$$
$$n = 0,1,2,...$$

Constructive interference

$$d\sin\theta = n\lambda$$
$$n = 0,1,2,...$$

2.8.2 Fringe spacing

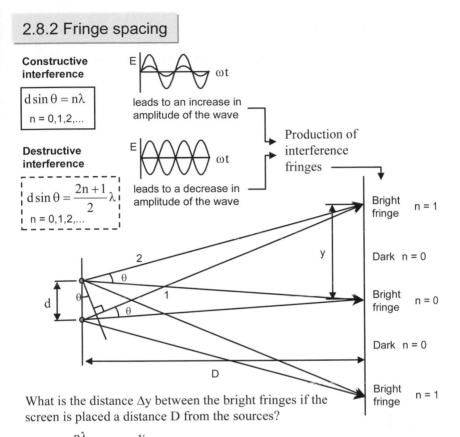

Constructive interference

$$d \sin \theta = n\lambda$$
$$n = 0,1,2,...$$

E   ⌠⌡ ωt

leads to an increase in amplitude of the wave

Destructive interference

$$d \sin \theta = \frac{2n+1}{2}\lambda$$
$$n = 0,1,2,...$$

E   ωt

leads to a decrease in amplitude of the wave

Production of interference fringes

Bright fringe   n = 1

Dark n = 0

Bright fringe   n = 0

Dark n = 0

Bright fringe   n = 1

2

1

θ

d

θ

θ

y

D

What is the distance Δy between the bright fringes if the screen is placed a distance D from the sources?

$$\sin \theta = \frac{n\lambda}{d}; \tan \theta = \frac{y}{D}$$

$$\frac{n\lambda}{d} = \frac{y}{D}$$

For small θ in radians, $\sin \theta = \tan \theta$

$$y = \frac{n\lambda D}{d}$$

distance from centre to nth bright fringe

$$y = (n+1)\frac{\lambda D}{d}$$

for the next adjacent fringe

The distance between any two fringes is thus:

$$\Delta y = \frac{\lambda D}{d}$$

The angular separation is $\tan \theta$:

$$\Delta \theta = \frac{\lambda}{d}$$

Note: these formulas apply for large value of D compared to d.

Intensity I is the amount of power per square metre. For an E field, the average intensity is given by:

$$I_{av} = \varepsilon_o c \frac{1}{2} E_o^{\,2}$$

Note that the intensity is proportional to the square of the amplitude E_o.

2.8.3 Fringe intensity

For two waves with the same amplitude E_o, then the interference maxima will have an amplitude $2E_o$ and the interference minima will have zero amplitude. What is the amplitude (and intensity) of the interference fringe pattern as a function of angle θ?

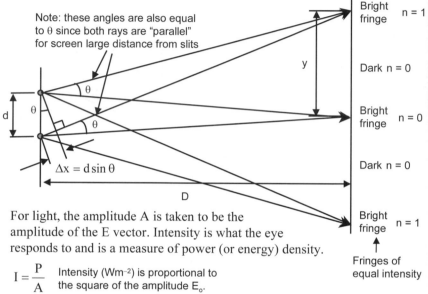

Note: these angles are also equal to θ since both rays are "parallel" for screen large distance from slits

$\Delta x = d \sin \theta$

Bright fringe n = 1

Dark n = 0

Bright fringe n = 0

Dark n = 0

Bright fringe n = 1

Fringes of equal intensity

For light, the amplitude A is taken to be the amplitude of the E vector. Intensity is what the eye responds to and is a measure of power (or energy) density.

$$I = \frac{P}{A}$$ Intensity (Wm^{-2}) is proportional to the square of the amplitude E_o.

If the two waves of amplitude E_o arrive at the screen out of phase by an angle $\Delta\phi$, then it can be shown that the resultant amplitude is:

$$E_R = 2E_o \cos\frac{\Delta\phi}{2}$$ It can be seen for $\Delta\phi = 0$, the maximum amplitude of the resultant E_{Rmax} is twice that of each component wave E_o and for minima, $\Delta\phi = \pi$ and the amplitude is zero.

The intensity is proportional to the square of the amplitude. Thus, the variation of intensity with θ is given by:

$$I = 4I_o \cos^2\frac{\Delta\phi}{2} \longrightarrow \Delta\phi = \frac{2\pi}{\lambda}d\sin\theta \quad \Delta\phi \text{ is in radians}$$

$$= 4I_o \cos^2\left(\frac{\pi}{\lambda}d\sin\theta\right)$$

$$I = I_{max} \cos^2\left(\frac{\pi}{\lambda}d\sin\theta\right)$$

Note: This formula actually says that the peaks all have the same intensity since $d\sin\theta = n\lambda$. In practice, this is not the case due to diffraction.

I_{max} is the maximum intensity of the resultant interference pattern. I_o is the maximum intensity of each individual component wave.

2.8.4 Thin film interference

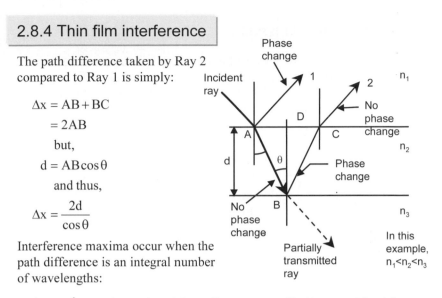

The path difference taken by Ray 2 compared to Ray 1 is simply:

$$\Delta x = AB + BC$$

$$= 2AB$$

but,

$$d = AB\cos\theta$$

and thus,

$$\Delta x = \frac{2d}{\cos\theta}$$

Interference maxima occur when the path difference is an integral number of wavelengths:

$$\Delta x = m\lambda$$
$$= \frac{2d}{\cos\theta}$$

where m is an integer. For near normal incidence, cos θ = 1 (i.e. θ = 0). However, one must be very careful about changes of phases at interfaces since this will influence the condition for constructive or destructive interference.

For a film with a thickness d, and light at normal incidence, **interference maxima** occur when:

$$\Delta x = m\lambda$$
$$= 2d$$

m = 0,1,2...

This applies if neither or both have a half-cycle phase shift on reflection. If one or the other rays undergoes a phase shift, then this becomes the condition for destructive interference.

Interference minima

$$\Delta x = \left(m + \frac{1}{2} \right)\lambda$$
$$= 2d$$

This is the condition for destructive interference if neither or both waves half a half-cycle reflection phase shift. If one or the other rays do have such a phase shift, then this becomes condition for constructive interference.

For a wave travelling from material n_1 to n_2, a phase change of π occurs **on reflection** if $n_1 < n_2$. No phase change when $n_1 > n_2$.

2.8.5 Optical path length

When light travels through a medium, it does so at a lower velocity (v) compared to that in a vacuum (c).

$$v = \frac{c}{n}$$

Velocity of light in vacuum

Refractive index of medium

$$\text{Now } \frac{\text{distance}}{\text{time}} = v$$

If the distance is the thickness x of the medium, then:

$$= \frac{c}{n}$$

$$\frac{(n)(x)}{t} = c$$

When light waves strike "denser" medium, velocity and wavelength both decrease, frequency remains constant.

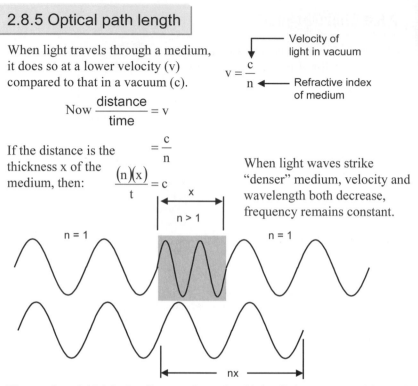

The product (n)(x) is the distance through which a light wave would travel if it were travelling with a speed c during time interval t. That is, since the velocity of the light in the medium (of thickness x) is v, which is less than c, then if the light wave were to travel through a vacuum with velocity c, then the distance travelled would be greater for the same time period t.

Light wave takes time t to traverse distance t through medium at velocity v

Light wave travels a greater distance, nt, in a vacuum with speed c in the same time t

---

(n)(x) = **optical path length** (o.p.l.)

Why bother with optical path length? Because when working out whether the path difference is an integral number of wavelengths (for constructive interference) we need to work out the *optical* path difference. That is, the path difference that two light waves would have if they were both travelling through a vacuum with speed c.

2.8.6 Examples

1. What colour does an oil film 300 nm thick appear to
    the eye when illuminated at normal incidence?

    Solution:

    $\Delta x = m\lambda$

    $\quad = 2d$

    $\lambda = 2\left(300 \times 10^{-9}\right)$

    $\quad = 600\,nm$

    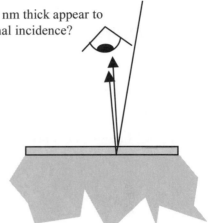

    Constructive interference at this
    wavelength, therefore the film
    will appear reddish

2. A double slit interferometer uses light of 620 nm. A thin film of
    transparent material (n = 1.6) is placed in the path of one of the beams.
    A shift of 50 bright fringes is observed. Determine the thickness of the
    film.

    Solution:

    Now, with film in position,
    the change in optical path
    length is $\Delta x = (nx - x)$ and
    this corresponds to 50
    fringes. Thus, $\Delta x = 50\lambda$.

    $\Delta x = n_r x - x$

    $\quad = \left(n_r - 1\right)x$

    $\left(1.6 - 1\right)x = 50\lambda$

    $0.6x = 50\left(620 \times 10^{-9}\right)$

    $x = 51.5\,\mu m$

2.9 Diffraction

Summary

$$I = I_{max} \frac{\sin^2 \dfrac{\alpha}{2}}{\left(\dfrac{\alpha}{2}\right)^2}$$ Single slit diffraction pattern

$$\alpha = \frac{2\pi}{\lambda} a \sin \theta$$ Phase difference

$$I = I_{max} \cos^2\left(\frac{\pi}{\lambda} d \sin \theta\right) \left[\frac{\sin^2 \alpha/2}{(\alpha/2)^2}\right]$$ Double slit diffraction pattern

$$\sin \theta \approx \theta = 1.22 \frac{\lambda}{a}$$ Circular aperture diffraction pattern

2.9.1 Interference and diffraction

Constructive and/or destructive interference arises from the superposition of waves. Previously, we examined the interference pattern generated by *two* coherent point sources of light.

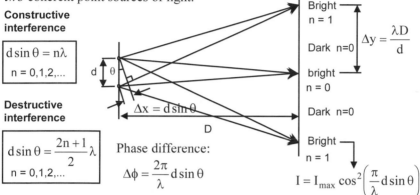

Constructive interference

$$\boxed{d\sin\theta = n\lambda}$$
$$n = 0,1,2,...$$

Destructive interference

$$\boxed{d\sin\theta = \frac{2n+1}{2}\lambda}$$
$$n = 0,1,2,...$$

Bright — n = 1

Dark n=0

bright — n = 0

Dark n=0

Bright — n = 1

$$\Delta y = \frac{\lambda D}{d}$$

$$\Delta x = d\sin\theta$$

Phase difference:

$$\Delta\phi = \frac{2\pi}{\lambda}d\sin\theta$$

$$I = I_{max}\cos^2\left(\frac{\pi}{\lambda}d\sin\theta\right)$$

The term **diffraction** describes the interference pattern arising from the superposition of an *infinite* number **point sources** of light waves and arises when the light waves interact with some obstruction in the light path. Diffraction effects become more pronounced when the dimensions of the object are comparable with the wavelength of the light.

A precise mathematical treatment of diffraction is very complicated. A simpler, but qualitative explanation may be found using **Huygen's principle**.

> A **plane wave** is equivalent to that produced by an infinite number of stationary point sources

Consider an infinite number of point sources arranged in a straight line:

r = vt

v

Spherical waves leave each source with velocity v. At a time t, the wave fronts have moved through a distance r = vt. Thus, circles of radius r give the position of the spherical wave front at a time t after leaving the source.

The envelope of these circles is a line parallel with the line of the sources (whether this line be straight or curved, it doesn't matter). In this example, the the envelope represents a plane wave.

2.9.3 Diffraction - circular aperture

Consider what happens when an obstruction is placed in
the path of the wave:

Waves tend to "bend" around the obstruction. What is the amplitude (and
hence intensity) of the distorted wave at a distant point P?

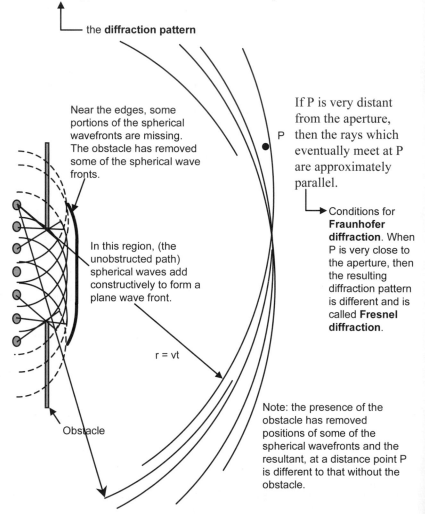

the **diffraction pattern**

Near the edges, some
portions of the spherical
wavefronts are missing.
The obstacle has removed
some of the spherical wave
fronts.

If P is very distant
from the aperture,
then the rays which
eventually meet at P
are approximately
parallel.

In this region, (the
unobstructed path)
spherical waves add
constructively to form a
plane wave front.

Conditions for
**Fraunhofer
diffraction**. When
P is very close to
the aperture, then
the resulting
diffraction pattern
is different and is
called **Fresnel
diffraction**.

$r = vt$

Obstacle

Note: the presence of the
obstacle has removed
positions of some of the
spherical wavefronts and the
resultant, at a distance point P
is different to that without the
obstacle.

2.9.4 Diffraction - single slit

Now consider the wave as it passes through the slit and beyond. Let the slit be represented by a total of N point sources within the slit.

At a large distance from slit, rays on the screen are virtually parallel (not shown parallel here) and arrive at P at some angle 90-θ.

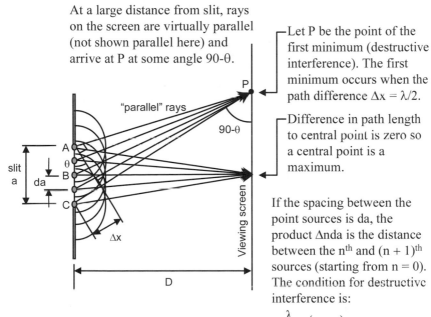

—Let P be the point of the first minimum (destructive interference). The first minimum occurs when the path difference Δx = λ/2.

—Difference in path length to central point is zero so a central point is a maximum.

If the spacing between the point sources is da, the product Δnda is the distance between the nth and (n + 1)th sources (starting from n = 0). The condition for destructive interference is:

$$\frac{\lambda}{2} = (\Delta nda)\sin\theta$$

If, for a "certain" θ, the waves from sources n = 0 and n = 2 satisfy this condition (Δn = 2), then, the intensity at point P will be diminished due to the destructive interference between these two rays. But, a ray leaving from, say n = 1 will not necessarily satisfy this condition hence there may be some light at P emanating from the source n = 1.

BUT, the ray leaving from n = 1 and going to P will cause destructive interference with that leaving from n = 3 (Δn = 2) since the path difference for this pair is also λ/2 (remember that the light rays are virtually parallel as they travel to P). Hence, the light from n = 1 is cancelled by that leaving from n = 3. This same argument applies to all of the point sources from A to C if there is always a pair of rays, separated by Δn da which results in complete destructive interference at P.

What is the value of θ for the first minimum?

As θ increases from 0, then it is evident that rays from sources at the extreme ends of the slit ($\Delta n = N$) will *first* meet the condition for destructive interference (i.e. their path length will first differ by $\lambda/2$, the path differences for all other sources being smaller). Thus, it may be at first thought that the condition $\lambda/2 = a\sin\theta$ (since $a = Nda$) will apply to the first minimum.

However, although these two rays will interfere destructively at P, the other rays will not since they are not paired with any corresponding ray with a path difference of $\lambda/2$. It is not until the source located at half way, "B" matches with the source at the edge "A" do sources in between have a matching ray at $\Delta x = \lambda/2$. Thus, the condition for the first absolute minimum is:

$$\frac{\lambda}{2} = \frac{a}{2}\sin\theta$$

The angular position of the first, and all subsequent, <u>minima</u> are found from:

$$\boxed{m\lambda = a\sin\theta} \quad m = 1,2,3..... \text{ but not } m = 0$$

slit width

the central
maximum

Note: as a decreases (slit gets narrower), the angle θ increases (pattern spreads out further). As slit gets wider (a increases) the angle θ decreases (conditions approach that of geometrical optics - no spreading out of rays).

This formula gives the direction of the minima (and hence the maxima) but says nothing about the intensity distribution in the diffraction pattern.

For the first minimum each side of the central maximum, the angle θ is generally small and thus, with $m = 1$, we have:

$$\frac{\lambda}{a} \approx \theta$$

Half-width of
central maximum.

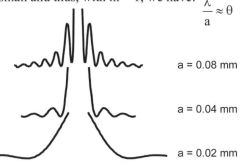

a = 0.08 mm

a = 0.04 mm

a = 0.02 mm

The interference maxima occur (only approximately) midway between the minima and the central maximum has twice the width of the others.

How to get the intensity of the diffraction pattern as a function of θ?

Each point within the slit acts as a source of waves. Proceeding down the slit, each wave will be progressively more out of phase than the source at the top of the slit until we reach the condition for the first minimum and the cycle of phase shifts repeats. Between each source, there is an equal phase difference $\Delta\alpha$. The <u>total</u> phase difference between the top and bottom waves is:

$$\alpha = \frac{2\pi}{\lambda}a\sin\theta$$

$\uparrow$ $2\pi/\lambda$ x path difference

Addition of individual E vectors, each with a phase difference $\Delta\alpha$, shows that the resultant amplitude variation, as a function of θ, is:

$$E = E_o \frac{\sin\frac{\alpha}{2}}{\frac{\alpha}{2}}$$

Note: this is different to the expression given previously for the addition of two waves of amplitude E_o and phase difference ϕ. Here we have an "infinite" number of point sources separated by a phase difference $\Delta\alpha$. The resultant amplitude E is expressed here in terms of the total phase difference a between the first source and the last.

Thus, since I is proportional to the amplitude E squared:

Note: This equation is indeterminate at $\alpha = 0$ (i.e. parallel rays from the slit to the central maximum $\theta = 0$). However, it can be shown (using calculus) that I converges to I_{max} at this condition.

$$I = I_{max} \frac{\sin^2\frac{\alpha}{2}}{\left(\frac{\alpha}{2}\right)^2}$$

$$\alpha = \frac{2\pi}{\lambda}a\sin\theta$$

Check on directions:

At what angle θ is intensity = 0? When $\alpha/2$ is a multiple of π.

$$\frac{\alpha}{2} = \frac{\pi}{\lambda}a\sin\theta$$

$$= m\pi$$

$$m\lambda = a\sin\theta \quad \text{as before}$$

At what θ is the intensity a maximum? When $\alpha/2$ is an odd multiple of $\pi/2$.

Only approximately correct. Presence of $\alpha/2$ in the denominator affects the value of I as well as the sine function in the numerator.

2.9.5 Diffraction - double slit

We have previously examined interference from two coherent point sources and saw that the condition for constructive interference was:

$$d \sin \theta = n\lambda \qquad n = 0, 1, 2, ...$$

and (without proving) that the intensity of the fringe pattern (which we now call the "diffraction pattern") is written:

$$I = I_{max} \cos^2\left(\frac{\pi}{\lambda} d \sin \theta\right)$$

Now, for a single slit, minima occur at: $m\lambda = a \sin \theta$

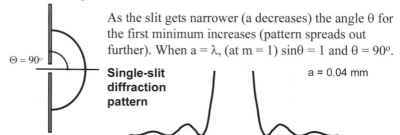

$\Theta = 90°$

As the slit gets narrower (a decreases) the angle θ for the first minimum increases (pattern spreads out further). When $a = \lambda$, (at m = 1) $\sin\theta = 1$ and $\theta = 90°$.

Single-slit diffraction pattern

a = 0.04 mm

The intensity distribution for two such slits is that previously calculated for two point sources separated by a distance d. However, if the slits are now made a little wider than λ, then the intensity distribution for the two point source condition is *modulated* by the single-slit diffraction pattern.

Double-slit diffraction pattern

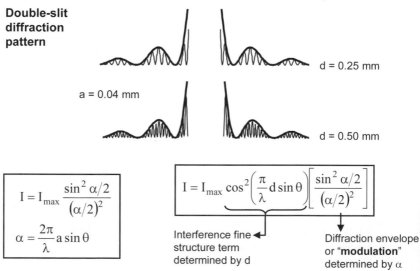

d = 0.25 mm

a = 0.04 mm

d = 0.50 mm

$$I = I_{max} \frac{\sin^2 \alpha/2}{(\alpha/2)^2}$$

$$\alpha = \frac{2\pi}{\lambda} a \sin \theta$$

$$I = I_{max} \underbrace{\cos^2\left(\frac{\pi}{\lambda} d \sin \theta\right)}_{} \underbrace{\left[\frac{\sin^2 \alpha/2}{(\alpha/2)^2}\right]}_{}$$

Interference fine structure term determined by d

Diffraction envelope or "**modulation**" determined by α

2.9.6 Diffraction - circular aperture

For plane waves arriving at a **circular aperture**, the resulting **far-field diffraction pattern** is similar to that obtained from a single slit but consists of circular, rather than straight, fringes. The angular separation between the the central bright spot and the first minimum is given by:

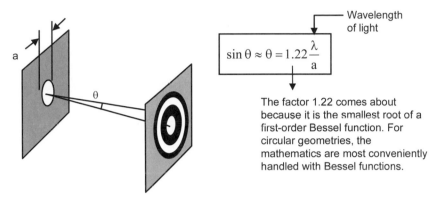

Wavelength of light

$$\sin \theta \approx \theta = 1.22 \frac{\lambda}{a}$$

The factor 1.22 comes about because it is the smallest root of a first-order Bessel function. For circular geometries, the mathematics are most conveniently handled with Bessel functions.

The entrance to the lens of the eye can be considered a circular aperture. If light from two point sources is incident on the eye, the diffraction pattern from each impinges on the retina. Experiment shows that the eye can sense a minimum change in intensity of about 20%. Thus, two sources can be resolved when the intensity between the maxima falls to within about 80% of the peak intensity.

This condition is met when the central maximum of one pattern coincides with the first minimum of the second. This is called the **Rayleigh criterion**. If $d\theta$ is the angle subtended at the eye by the two sources, then the **resolving power**, defined as $1/\theta$, can be calculated from:

$$a \sin \theta = 1.22\lambda$$
$$\approx a\theta \quad \text{since } \theta \text{ is}$$
$$\approx a \frac{d}{D} \quad \text{usually small}$$

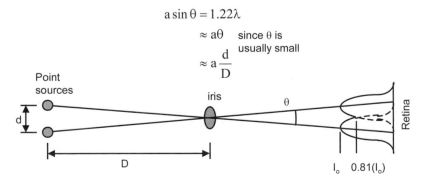

2.9.7 Example

1. A microscope has an objective lens of diameter 4 mm and a focal length of 3.2 mm. Calculate the minimum distance between two objects which may just be resolved using this lens if the objects are illuminated with light of wavelength (a) 500 nm and (b) 620 nm and if the object is placed at the focal point of the lens.

Solution:

$$f = 0.0032\text{mm}$$

$$a = 0.004$$

$$\lambda = 500\text{nm}$$

$$a\sin\theta = 1.22\lambda$$

(a) $$0.004\sin\theta = 500\times10^{-9}1.22$$

$$\sin\theta = 1.525\times10^{-4}$$

$$= \frac{d}{D}$$

$$d = 0.0032\left(1.525\times10^{-4}\right)$$

$$= 488\,\text{nm}$$

$$\lambda = 620\,\text{nm}$$

(b) $$0.004\sin\theta = 1.22\left(620\times10^{-9}\right)$$

$$\sin\theta = 1.89\times10^{-4}$$

$$d = 0.0032\left(1.89\times10^{-4}\right)$$

$$= 605\,\text{nm}$$

2.10 Polarisation

Summary

$$\tan \theta_p = \frac{n_2}{n_1}$$ Brewster's law

$$I = I_o \cos^2 \theta$$ Malus' law

2.10.1 Polarised light

Light waves are transverse waves. The varying E and B fields have a direction which is at right angles (ie. transverse) to the direction of propagation of the wave. The transverse nature of light waves lead to some very interesting physical effects. The term **polarisation** is the transverse property of the light wave which leads to these effects.

For a "single" ray of light, if the direction of the E vectors remains constant, then the light is said to be **polarised** in this direction.

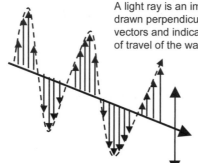

A light ray is an imaginary line drawn perpendicular to the E vectors and indicates the direction of travel of the waves.

Since the most common effects of light are due to the action of the electric field E, we shall only show the direction of E in these diagrams to avoid clutter. At all times, an accompanying B field is assumed to be also present.

It is often convenient to resolve the E vectors in a light ray into horizontal and vertical components. For **linearly polarised light** (θ is a constant) of constant intensity, the magnitude of the components remain constant

When the direction of polarisation changes in a regular periodic manner, the light is said to be elliptically polarised. For **elliptically polarised light**, the magnitudes of the components change as the magnitude and direction of E changes. Circularly polarised light is just a special case of elliptically polarised light and occurs when the magnitude of the E vector remains constant but its direction changes.

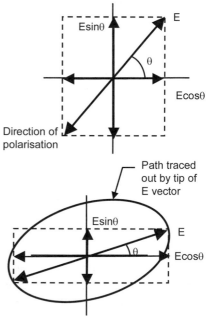

2.10.2 Unpolarised light

A more complete understanding of the nature of polarised light can be had when one considers the nature of **unpolarised light**.

Most light sources emit light as a result of de-excitation of atoms. Although the light emitted from one particular atom may be polarised at a particular direction, it is unlikely that the plane, or direction, of polarisation of any one atom will be the same as any other (except in a laser). Thus, most light consists of a collection of light rays which have all possible directions of polarisation. Further, the emission of light from one particular atom occurs over a very short time period.

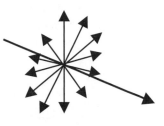

The significance of this is that in a ray of unpolarised light, the direction of E would be in one direction for an instant, and then another random direction in the next instant. In practice, the time intervals between these changes in direction of polarisation are so small that the light appears to consist of E fields of all possible directions at once.

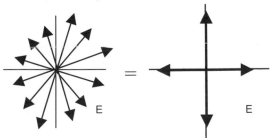

For a beam of unpolarised light, all the momentary polarised rays which make up the beam can all be resolved into a single vertical and a single horizontal component - *which are equal in magnitude at any one time* due to the randomness of the **atomic oscillators** which produce the waves in the first place - and the very short time in which these rays appear after one another in the beam.

The magnitude of the components change rapidly and irregularly. Further, unpolarised light does not consist of *continuously* varying sinusoidal E fields. Thus, the phase of the two components varies randomly (but together) in time.

2.10.3 Sources of polarised light

Consider a beam of unpolarised light passing through a sheet of **polarising** material. The light can be resolved into vertical and horizontal components of equal magnitude. The component that is aligned with the molecules in the material are absorbed, the other component is transmitted.

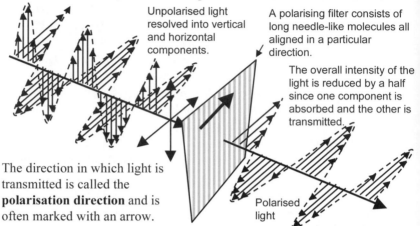

Unpolarised light resolved into vertical and horizontal components.

A polarising filter consists of long needle-like molecules all aligned in a particular direction.

The overall intensity of the light is reduced by a half since one component is absorbed and the other is transmitted.

The direction in which light is transmitted is called the **polarisation direction** and is often marked with an arrow.

Polarised light

Now consider **reflection** from a surface. The incoming E vector causes charged particles within the atoms of the surface to oscillate resulting in radiation emission. Now, when the incoming light ray strikes an optically more dense material ($n_2 > n_1$) at an angle θ_p the **polarising angle**, E vectors parallel to the plane of incidence are absorbed and refracted but not reflected. Only the E vectors perpendicular to the plane of incidence are reflected. The reflected ray is polarised in a direction perpendicular to the plane of incidence. The polarising angle occurs when the angle between the reflected and refracted ray is 90°.

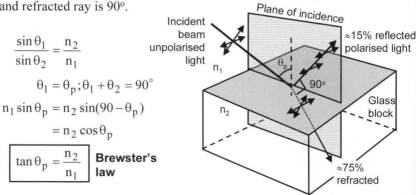

Plane of incidence

Incident beam unpolarised light n_1

≈15% reflected polarised light

θ_p

90°

Glass block

n_2

≈75% refracted

$$\frac{\sin\theta_1}{\sin\theta_2} = \frac{n_2}{n_1}$$

$$\theta_1 = \theta_p ; \theta_1 + \theta_2 = 90°$$

$$n_1 \sin\theta_p = n_2 \sin(90 - \theta_p)$$

$$= n_2 \cos\theta_p$$

$$\boxed{\tan\theta_p = \frac{n_2}{n_1}}$$ **Brewster's law**

2.10.4 Malus' law

Unpolarised light is incident on a polarising material and polarised light
of intensity half of that of the unpolarised light is transmitted. If this
polarised light is then incident on a second sheet of polarising material,
then the intensity of the beam transmitted through the second sheet
depends on the angle of the second polariser with respect to the first.

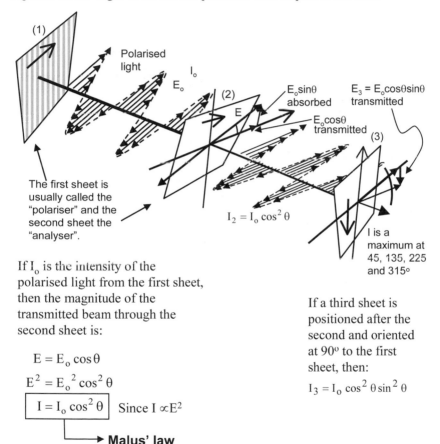

The first sheet is
usually called the
"polariser" and the
second sheet the
"analyser".

If I_o is the intensity of the
polarised light from the first sheet,
then the magnitude of the
transmitted beam through the
second sheet is:

$$E = E_o \cos \theta$$

$$E^2 = E_o^2 \cos^2 \theta$$

$$\boxed{I = I_o \cos^2 \theta} \quad \text{Since } I \propto E^2$$

$\longrightarrow$ **Malus' law**

The intensity of the transmitted beam is a
maximum at $\theta = 0$ and zero at $\theta = 90°$.

If a third sheet is
positioned after the
second and oriented
at 90° to the first
sheet, then:

$$I_3 = I_o \cos^2 \theta \sin^2 \theta$$

2.10.5 Example

1. Polaroid sunglasses work by reducing glare associated with polarised
    light reflecting from surfaces. For incident sunlight on a pond,
    determine (a) the angle at which the reflected light is completely
    polarised, and (b), the reduction in intensity of the reflected light for a
    person wearing sunglasses who sees the reflected light where the
    polarised material is at an angle of 30° to the direction of polarisation
    of the light.

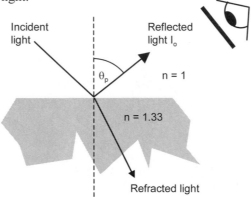

Solution:

(a) From Brewster's law:

$$\tan\theta_p = \frac{n_2}{n_1}$$

$$\theta_p = \tan^{-1}\frac{1.33}{1}$$

$$= 53.1°$$

(a) From Malus' law:

$$I = I_o \cos^2\theta$$

$$\frac{I}{I_o} = \cos^2 30$$

$$= 75\% \quad \text{i.e. 25\% reduction in intensity}$$

Part 3

Electricity

3.1 Electricity

Summary

$$F = k\frac{q_1 q_2}{d^2}$$ Force between two charges

$$F = q_1 E$$ Force on a charge in a field

$$E = 4\pi k \frac{Q}{A}$$ Electric field - point charge

$$\overline{E} = k\frac{q\hat{r}}{r^2}$$ Electric field - point charge

$$\phi = EA$$ Electric flux

$$I = A\left(q_1 n_1 v_1 + (-q_2)n_2(-v_2)\right)$$

$$i = \frac{dq}{dt}$$ Electric current

$$\frac{W}{q} = Ed$$ Electric potential

$$\frac{V}{I} = R$$ Ohm's law

$$P = VI = I^2 R$$ Power - resistor

$$R = \rho \frac{l}{A}$$ Resisitivity

$$C = \frac{Q}{V} = \varepsilon_0 \frac{A}{d}$$ Capacitance

$$L = \mu_0 A \frac{N^2}{l}$$ Inductance

$$U = \frac{1}{2}CV^2$$ Energy - capacitor

$$U = \frac{1}{2}LI^2$$ Energy - inductor

$$R_{AB} = R_1 + R_2$$ Resistors- series

$$\frac{1}{R_{AB}} = \frac{1}{R_1} + \frac{1}{R_2}$$ Resistors - parallel

$$R_{AB} = \frac{R_1 R_2}{R_1 + R_2}$$

3.1.1 Electric charge

Electrical (and magnetic) effects are a consequence of a property of matter called **electric charge**. Experiments show that there are two types of charge that we label **positive** and **negative**. Experiments also show that unlike charges attract and like charges repel.

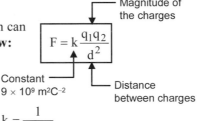

The charge on a body usually refers to its *excess* or net charge. The smallest unit of charge is that on one electron e $= 1.60219 \times 10^{-19}$ Coulombs.

The force of attraction or repulsion can be calculated using **Coulomb's law:**

- **Repulsive forces** between like charges are negative
- **Attractive forces** between unlike charge are positive

$$F = k\frac{q_1 q_2}{d^2}$$

Magnitude of the charges

Constant 9×10^9 m²C⁻²

Distance between charges

If the two charges are in some substance, eg. Air, then the **Coulomb force** is reduced. Instead of using ε_o, we must use ε for the substance. Often, the **relative permittivity** ε_r is specified.

$$k = \frac{1}{4\pi\varepsilon_o}$$ → the **permittivity of free space** $= 8.85 \times 10^{-12}$ Farads/metre

$$\varepsilon_r = \frac{\varepsilon}{\varepsilon_o}$$

Material	ε_r
Vacuum	1
Water	80
Glass	8

1. Imagine that one of the charges is hidden from view.

2. The other charge still experiences the Coulomb force and thus we say it is acted upon by an **electric field**.

3. If a **test charge** experiences a force when placed in a certain place, then an electric field exists at that place. The direction of the field is taken to be that in which a positive test charge would move in the field.

$$F = k\frac{q_1 q_2}{d^2}$$

$$\text{Let } E = k\frac{q_2}{d^2}$$

$$\text{Thus } F = q_1 E$$

Note: The origin of the field E may be due to the presence of many charges but the magnitude and direction of the resultant field E can be obtained by measuring the force F on a single test charge q.

3.1.2 Electric flux

An electric field may be represented by lines of force. The total number of lines is called the **electric flux**. The number of lines per unit cross-sectional area is the **electric field intensity**, or simply, the magnitude of the electric field.

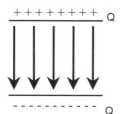

+ + + + + + + + Q

Q

Uniform electric field between two charged parallel plates

$$E = 4\pi k \frac{Q}{A}$$

Non-uniform field surrounding a point charge

$$\overline{E} = k \frac{q\hat{r}}{r^2}$$

- Arrows point in direction of path taken by a positive test charge placed in the field
- Number density of lines crossing an area A indicates electric field intensity
- Lines of force start from a positive charge and always terminate on a negative charge (even for an **isolated charge** where the corresponding negative charge may be quite some distance away).

Note: For an isolated charge (or charged object) the termination charge is so far away that it contributes little to the field. When the two charges are close together, such as in the parallel plates, both positive and negative charges contribute to the strength of the field. For the plates, Q is the charge on either plate, a factor of 2 has already been included in the formula.

How to calculate electric flux (e.g. around a point charge)

$$A = 4\pi R^2 \quad \text{Area of a sphere radius R}$$

$$E \propto \frac{N}{A} \quad \text{by definition}$$

But $EA \propto N$ electric flux

Thus $E = \dfrac{kq}{R^2}$ $k = 1/4\pi\varepsilon_o$

$$EA = \frac{kq}{R^2} 4\pi R^2$$

$$= 4\pi kq \quad \text{independent of R but proportional to N}$$

$$= \frac{q}{\varepsilon_o}$$

$$= \phi \quad \text{Electric flux}$$

3.1.3 Conductors and insulators

Atoms consist of a positively charged **nucleus** surrounded by negatively charged **electrons**. Solids consist of a fixed arrangement of atoms usually arranged in a lattice. The position of individual atoms within a solid remains constant because **chemical bonds** hold the atoms in place. The behaviour of the outer electrons of atoms are responsible for the formation of chemical bonds. These outer shell electrons are called **valence electrons**.

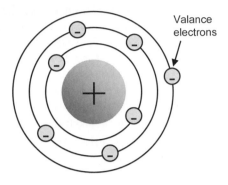

Valance electrons

1. Conductors

Valence electrons are weakly bound to the atomic lattice and are free to move about from atom to atom.

2. Insulators

Valence electrons tightly bound to the atomic lattice and are fixed in position.

3. Semiconductors

In **semiconductors**, valence electrons within the crystal structure of the material not as strongly bound to the atomic lattice and if given enough energy, may become mobile and free to move just like in a conductor.

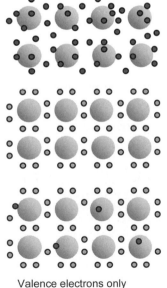

Valence electrons only shown in these figures

Electrons, especially in conductors, are **mobile charge carriers** (they have a charge, and they are mobile within the atomic lattice).

3.1.4 Electric current

Mobile charge carriers may be either positively charged (e.g. positive ions in solution), negatively charged (e.g. negative ions, loosely bound valence electrons). Consider the movement during a time Δt of positive and negative charge carriers in a **conductor** of cross-sectional area A and length l placed in an electric field E:

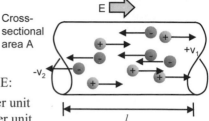

Cross-sectional area A

Let there be n_1 positive carriers per unit volume and n_2 negative carriers per unit volume. Charge carriers move with drift velocities v_1 and $-v_2$. In time Δt, each particle moves a distance $l = v_1\Delta t$ and $l = v_2\Delta t$.

The total positive charge exiting from the right (and entering from the left) during Δt is thus:

$$Q^+ = q_1 n_1 (v_1\Delta t)A$$

Total charge Coulombs

Charge on one mobile carrier

No. of charge carriers per unit volume

Volume

The total negative charge exiting from the left is $Q^- = n_2(-q_2)(-v_2\Delta t)A$. The total net movement of charge during Δt is thus:

$$Q^+ + Q^- = q_1 n_1 v_1 A\Delta t + (-q_2)n_2(-v_2)A\Delta t$$

The total charge passing any given point in Coulombs per second is called electric current:

$$\frac{Q^+ + Q^-}{\Delta t} = q_1 n_1 v_1 A + (-q_2)n_2(-v_2)A$$

1 Amp is the rate of flow of electric charge when one Coulomb of electric charge passes a given point in an electric circuit in one second.

$$I = A\big(q_1 n_1 v_1 + (-q_2)n_2(-v_2)\big)$$

Current density $J = I/A$
Amps m^{-2} or
Coulombs $m^{-2}\,s^{-1}$

In general,

$$i = \frac{dq}{dt}$$

Lower case quantities refer to instantaneous values. Upper case refers to steady-state or DC values.

In metallic conductors, the mobile charge carriers are negatively charged electrons hence $n_1 = 0$.

Note that the **Amp** is a measure of quantity of charge per second and provides no information about the net drift <u>velocity</u> of the charge carriers (≈ 0.1 mm s^{-1}).

3.1.5 Conventional current

Electric current involves the net flow of **electrical charge carriers**, which, in a metallic conductor, are negatively charged electrons. Often in circuit analysis, the physical nature of the actual flow of charge is not important - whether it be the flow of free electrons or the movement of positive ions in a solution.

This happens due to the movement of both positive and negative ions within the solution inside the cell. →

But, in the 1830s, no one had heard of the **electron**. At that time, Faraday noticed that when current flowed through a wire connected to a chemical cell, the anode (positive) lost weight and the cathode (negative) gained weight, hence it was concluded that charge carriers flowed through the wire from positive to negative.

We now know that positive and negative ions in the solution move towards the anode and cathode in opposite directions and electrons in the wire go from negative to positive.

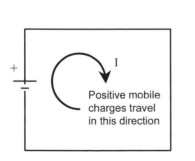

Positive mobile charges travel in this direction

There are two types of charge carriers: positive and negative. For historical reasons, all laws and rules for electric circuits are based on the direction that would be taken by positive charge carriers. Thus, in all circuit analysis, *imagine* that current flows due to the motion of positive charge carriers. Current then travels from positive to negative. This is called **conventional current**.

If we need to refer to the actual *physical process* of conduction, then we refer to the specific charge carriers appropriate to the conductor being considered.

3.1.6 Potential difference

When a charge carrier is moved from one point to another in an electric field, its potential energy is changed since this movement involved a force F moving through a distance.

If q+ is moved *against* the field, then work is done *on* the charge and the **potential energy** of the charge is increased.

$$F = qE$$

The work done is thus:

$$W = Fd$$
$$= qEd \quad \text{Joules}$$

The work per unit charge is called the **electrical potential**:

$$\boxed{\frac{W}{q} = Ed}$$ Joules/   or
                              Coulomb   Newtons/
                              (**Volt**)   Coulomb

If a charged particle is released in the field, then work is done on the particle and it acquires kinetic energy. The force acting on the particle is proportional to the field strength E. The stronger the field, the larger the force – the greater the acceleration and the greater the rate at which the charged particle acquires **kinetic energy**.

A uniform electric field E exists between two parallel charged plates since a positive test charge placed anywhere within this region will experience a downwards force of uniform value. The electric field also represents a **potential gradient**.

If the negative side of the circuit is grounded, then the electrical potential at the negative plate is zero and increases uniformly through the space between the plates to the top plate where it is +V.

The potential gradient (in volts per metre) is numerically equal to the **electric field strength** (Newtons per Coulomb) but is opposite in direction.

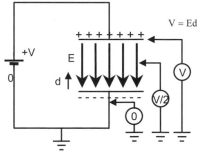

In a uniform electric field, the potential decreases uniformly along the field lines and is a potential gradient.

3.1.7 Resistance

A voltage source, utilising chemical or mechanical means, raises the electrical potential of mobile charge carriers (usually electrons) within it. There is a net build up of charge at the terminals of a voltage source. This net charge results in an **electric field** which is channelled through the conductor. Mobile electrons within the conductor thus experience an electric force and are accelerated.

However, as soon as these electrons move through the conductor, they suffer collisions with other electrons and fixed atoms and lose velocity and thus some of their kinetic energy. Some of the fixed atoms correspondingly acquire **internal energy** (vibrational motion) and the **temperature** of the conductor rises. After collision, electrons are accelerated once more and again suffer more collisions.

Note: Negatively charged electrons move in opposite direction to that of electric field.

Alternate accelerations and decelerations result in a net average velocity of the mobile electrons (called the **drift velocity**) which constitutes an electric current. **Electrical potential energy** is converted into **heat** within the **conductor**. The decelerations arising from collisions is called **resistance**.

Experiments show that, for a particular specimen of material, when the applied voltage is increased, the current increases. For most materials, doubling the voltage results in a doubling of the current. That is, the current is directly proportional to the current: $I \propto V$

The constant of proportionality is called the **resistance**. Resistance limits the current flow through a material for a particular applied voltage.

$$\frac{V}{I} = R \qquad \textbf{Ohm's law}$$

Units: **Ohms**

The rate at which electrical potential energy is converted into heat is the **power** dissipated by the resistor. Since electrical potential is Joules/Coulomb, and current is measured in Coulombs/second, then the product of voltage and current gives Joules/second which is **power** (in **Watts**).

$$P = VI$$

but $\quad V = IR$

thus $\quad \boxed{P = I^2 R}$

3.1.8 Resistivity

Experiments show that the **resistance** of a particular specimen of material (at a constant temperature) depends on three things:

- the length of the conductor, l
- the cross-sectional area of the conductor, A
- the type of material, ρ

The material property which characterises the ability for a particular material to conduct electricity is called the **resistivity** ρ (the inverse of which is the **conductivity** σ).

Material	ρ Ωm @ 20 °C
Silver	1.64×10^{-8}
Copper	1.72×10^{-8}
Aluminium	2.83×10^{-8}
Tungsten	5.5×10^{-8}

The resistance R (in ohms) of a particular length l of material of cross-sectional area A is given by:

$$R = \rho \frac{l}{A}$$

The units of ρ are Ωm, the units of σ are S m^{-1}.

Now, $V = IR$

hence $\quad = I \dfrac{\rho l}{A}$

$\dfrac{V}{l} = \rho \dfrac{I}{A}$

The quantity I/A is called the **current density** J.

but $\quad E = \dfrac{V}{l}$

thus $\quad E = \rho \dfrac{I}{A}$ ◄——┐

The **number density** of **mobile charge carriers** n depends on the material. If the number density is large, then, if E (and hence v) is held constant, the **resistivity** must be small. Thus, the resistivity depends inversely on n. **Insulators** have a high resistivity since n is very small. **Conductors** have a low resistivity because n is very large.

$= \rho \dfrac{\sum nqvA}{A}$

Sum of the positive and negative mobile charge carriers
$\Sigma nqvA = n_1 q_1 v_1 + n_2(-q_2)(-v_2)A$

$$\rho = \frac{E}{nqv}$$

For a metal, only one type of mobile charge carrier

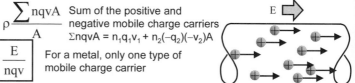

For a particular specimen of material, n, q, A and l are a constant. Increasing the applied field E results in an increase in the **drift velocity** v and hence an increase in current I.

The **resistivity** of a pure substance is lower than that of one containing impurities because the mobile electrons are more likely to travel further and acquire a larger velocity when there is a regular array of stationary atoms in the conductor.

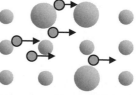

Presence of impurity atoms decrease the average drift velocity.

3.1.9 Variation of resistance

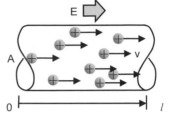

Consider an applied voltage
which generates an electric field
E within a conductor of resistance
R and of length l and area A.

1. Variation with area

Evidently, if the area A is increased, there will be more mobile charge
carriers available to move past a given point during a time Δt under the
influence of the field and the current I increases. Thus, for a particular
specimen, the resistance decreases with increasing cross-sectional area.

2. Variation with length

Now, the field E acts over a length l.

$$V = El$$

If the applied voltage is kept
constant, then it is evident that if l is
increased, E must decrease. The
drift velocity depends on E so that if
E decreases, then so does v, and
hence so does the current.

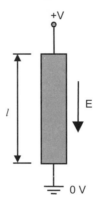

3. Variation with temperature

Increasing the **temperature** increases
the random thermal motion of the atoms in the conductor thus increasing
the chance of collision with a mobile electron thus reducing the average
drift velocity and increasing the resistivity. Different materials respond to
temperature according to the **temperature coefficient of resistivity** α.

$$\rho_T = \rho_0\left[1+\alpha(T-T_o)\right]$$
$$R_T = R_o\left[1+\alpha(T-T_o)\right]$$

R_T = resistance at T
R_o = resistance at T_o (usually 0 °C)

This formula applies to
a conductor, not a
semi-conductor.

Material	α (C^{-1} Ω^{-1})
Tungsten	4.5×10^{-3}
Platinum	3.0×10^{-3}
Copper	3.9×10^{-3}

3.1.10 Emf

Consider a **chemical cell**:

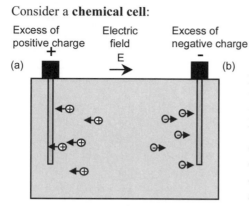

Excess of positive charge (a)

Electric field E

Excess of negative charge (b)

rather than "electrostatic"

Chemical attractions cause positive charges to build up at (a) and negative charges at (b)

Electrostatic repulsion due to build-up of positive charge at (a) eventually becomes equal to the chemical attractions tending to deposit more positive charges and system reaches equilibrium.

The term **Emf** (electromotive force) is defined as the amount of energy expended by the cell in moving 1 Coulomb of charge from (b) to (a) *within* the cell.

At **open circuit**, Emf = V_{ab}

"force" is poor choice of words since Emf is really "energy" (Joules per Coulomb)

Now connect an external load R_L across (a) and (b). The **terminal voltage** V_{ab} is now reduced.

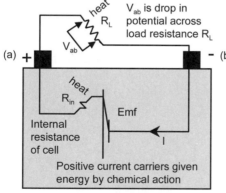

heat R_L

V_{ab} is drop in potential across load resistance R_L

(a) +   − (b)

heat R_{in}

Internal resistance of cell

Emf

I

Positive current carriers given energy by chemical action

Loss of positive charge from (a) reduces the accumulated charge at (a) and hence chemical reactions proceed and more positive charges are shifted from (b) to (a) within the cell to make up for those leaving through the external circuit. Thus, there is a steady flow of positive charge through the cell and through the wire.

Assume positive carriers – **conventional current** flow.

The circuit has been drawn to emphasize where potential drops and rises occur.

But, the continuous conversion of chemical potential energy to electrical energy is not 100% efficient. Charge moving within the cell encounters **internal resistance** which, in the presence of a current I, means a voltage drop so that:

At "closed circuit", Emf = $V_{ab} + IR_{in}$

3.1.11 Capacitance

Consider two **parallel plates** across which is placed a voltage V.

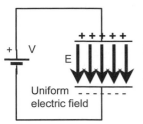

When a voltage V is connected across the plates, current begins to flow as charge builds up on each plate. In the diagram, **negative charge** builds up on the lower plate and **positive charge** on the upper plate. The accumulated charge on the two plates establishes an electric field between them. Since there is an electric field between the plates, there is an **electrical potential difference** between them.

For a **point charge** in space, E depends on the distance away from the charge:

$$E = \frac{1}{4\pi\varepsilon_0}\frac{q}{d^2}$$

Permittivity of free space = 8.85×10^{-12} Farads m^{-1}

But, for **parallel plates** holding a total charge Q on each plate, calculations show that the electric field E in the region between the plates is proportional to the magnitude of the charge Q and inversely proportional to the area A of the plates. For a given accumulated charge +Q and −Q on each plate, the field E is independent of the distance between the plates.

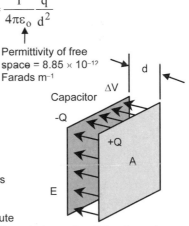

$$E = \frac{Q}{\varepsilon_0 A}$$

Q in these formulas refers to the charge on ONE plate. Both positive and negative charges contribute to the field E. A factor of 2 has already been included in these formulas.

Now, $V = Ed$

thus $V = \dfrac{Q}{\varepsilon_0 A}d$

A charged particle released between the plates will experience an accelerating force.

Capacitance is defined as the ratio of the magnitude of the charge on each plate (+Q or −Q) to the potential difference between them.

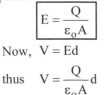

A large capacitor will store more charge for every volt across it than a small capacitor.

$$C = \frac{Q}{V} \quad \text{but} \quad V = \frac{Q}{\varepsilon_0 A}d$$

$$= Q\frac{\varepsilon_0 A}{Qd}$$

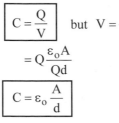

$$C = \varepsilon_0\frac{A}{d}$$

Units: **Farads**

If the space between the plates is filled with a **dielectric**, then capacitance is increased by a factor ε_r. A dielectric is an insulator whose atoms become polarised in the electric field. This adds to the **storage capacity** of the capacitor.

3.1.12 Capacitors

If a capacitor is charged and the voltage source V is then disconnected from it, the accumulated charge remains on the plates of the capacitor.

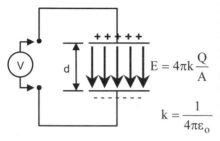

Since the charges on each plate are opposite, there is an **electrostatic force of attraction** between them but the charges are kept apart by the gap between the plates. In this condition, the capacitor is said to be **charged**. A **voltmeter** placed across the terminals would read the voltage V used to charge the capacitor.

$$E = 4\pi k \frac{Q}{A}$$

$$k = \frac{1}{4\pi\varepsilon_0}$$

When a **dielectric** is inserted in a capacitor, the molecules of the dielectric align themselves with the applied field. This alignment causes a field of opposite sign to exist within the material thus reducing the overall net field. For a given applied voltage, the total net field within the material is small for a material with a high **permittivity** ε. The permittivity is thus a measure of how easily the charges within a material line up in the presence of an applied external field.

If the plates are separated by a dielectric, then the field E is reduced. Instead of using ε_o, we must use ε for the substance. Often, the **relative permittivity** ε_r is specified.

In a conductor, charge carriers not only align but actually move under the influence of an applied field. This movement of charge carriers completely cancels the external field all together. The net electric field within a conductor placed in an external electric field is zero!

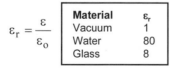

$$\varepsilon_r = \frac{\varepsilon}{\varepsilon_0}$$

Material	ε_r
Vacuum	1
Water	80
Glass	8

When a capacitor is connected across a voltage source, the current in the circuit is initially very large and then decreases as the capacitor charges. The voltage across the capacitor is initially zero and then rises as the capacitor charges.

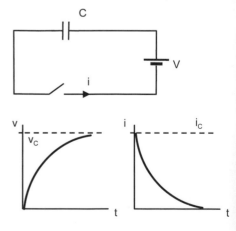

3.1.13 Inductance

In a conductor carrying a steady electric current, there is a magnetic field around the conductor. The magnetic field of a current-carrying conductor may be concentrated by winding the conductor around a tube to form a **solenoid**.

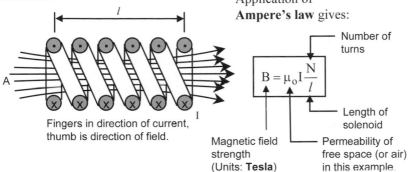

Application of **Ampere's law** gives:

$$B = \mu_o I \frac{N}{l}$$

- Number of turns
- Length of solenoid
- Permeability of free space (or air) in this example.
- Magnetic field strength (Units: **Tesla**)

Fingers in direction of current, thumb is direction of field.

When the current in the coil changes, the resulting change in **magnetic field** induces an emf in the coil (**Faraday's law**).

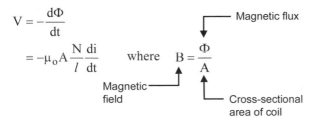

$$V = -\frac{d\Phi}{dt}$$

$$= -\mu_o A \frac{N}{l} \frac{di}{dt} \qquad \text{where} \qquad B = \frac{\Phi}{A}$$

- Magnetic field
- Magnetic flux
- Cross-sectional area of coil

But, this is the voltage induced in *each loop* of the coil. Each loop lies within a field B and experiences the changing current. The *total* voltage induced between the two ends of the coil is thus N times this:

$$V_{total} = -\mu_o A \frac{N^2}{l} \frac{di}{dt}$$

The induced voltage tends to oppose the change in current (**Lenz's law**).

Inductance: $\boxed{L = \mu_o A \frac{N^2}{l}}$ determines what voltage is induced within the coil for a given rate of change of current.

Units **Henrys**

3.1.14 Inductors

In a circuit with an inductor, when the **switch** closes, a changing current results in a changing magnetic field around the coil. This *changing* magnetic field induces a voltage (Emf) in the loops (Faraday's law) which tends to oppose the applied voltage (Lenz's law). Because of the self induced opposing Emf, the current in the circuit does not rise to its final value at the instant the circuit is closed, but grows at a rate which depends on the **inductance** (in **henrys**, L) and resistance (R) of the circuit. As the current increases, the *rate of change* of current decreases and the magnitude of the opposing voltage decreases. The current reaches a maximum value I when the opposing voltage drops to zero and all the voltage appears across the resistance R.

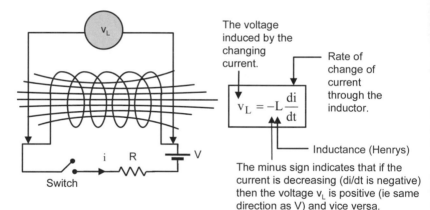

The voltage induced by the changing current.

Rate of change of current through the inductor.

$$v_L = -L\frac{di}{dt}$$

Inductance (Henrys)

The minus sign indicates that if the current is decreasing (di/dt is negative) then the voltage v_L is positive (ie same direction as V) and vice versa.

When the switch is closed, the rate of change of current is controlled by the value of L and R. Calculations show that the voltage across the inductor is given by $v_L = Ve^{-\frac{Rt}{L}}$

Switch is closed

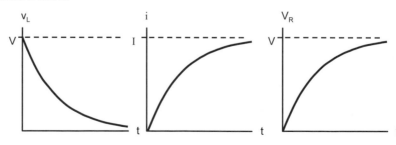

3.1.15 Energy and power

In a circuit with a **capacitor**, energy is expended by the voltage source as it forces charge onto the plates of the capacitor. When fully charged, and disconnected from the voltage source, the voltage across the capacitor remains. The stored electric potential energy within the charged capacitor may be released when desired by discharging the capacitor. To find the energy stored in a capacitor, we can start by considering the power used during charging it.

Power $P = vi$

$$i = \frac{dq}{dt}$$

$$Pdt = vdq = U$$

$$U = \int_0^Q vdq = \int_0^Q \frac{q}{C}dq \quad \text{Energy}$$

$$= \frac{1}{2}\frac{Q^2}{C}$$

$$\boxed{U = \frac{1}{2}CV^2}$$ Energy stored in a capacitor

Lower case letters refer to instantaneous quantities.

Establishing a current in an **inductor** requires energy which is stored in the **magnetic field**. When an inductor is discharged, this energy is released. The energy stored can be calculated from the power consumption of the circuit when a voltage is applied to it.

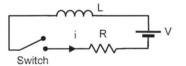

Switch

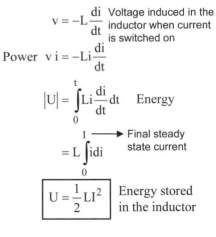

$$v = -L\frac{di}{dt}$$ Voltage induced in the inductor when current is switched on

Power $vi = -Li\frac{di}{dt}$

$$|U| = \int_0^t Li\frac{di}{dt}dt \quad \text{Energy}$$

$$= L\int_0^I idi$$

$I \longrightarrow$ Final steady state current

$$\boxed{U = \frac{1}{2}LI^2}$$ Energy stored in the inductor

The energy is released when the current decreases from I to 0 (i.e. when the circuit is broken.)

$$i = Ie^{-\frac{Rt}{L}}$$

$$v_R = Ve^{-\frac{Rt}{L}}$$

$$v_L = -Ve^{-\frac{Rt}{L}}$$

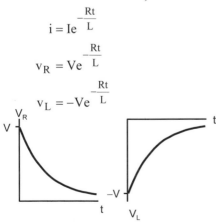

3.1.16 Circuits

Resistors in **series**

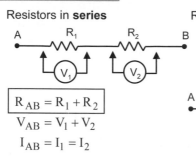

$$R_{AB} = R_1 + R_2$$
$$V_{AB} = V_1 + V_2$$
$$I_{AB} = I_1 = I_2$$

Resistors in **parallel**

$$\frac{1}{R_{AB}} = \frac{1}{R_1} + \frac{1}{R_2}$$

$$R_{AB} = \frac{R_1 R_2}{R_1 + R_2}$$

$$V_{AB} = V_1 = V_2$$
$$I_{AB} = I_1 + I_2$$

Capacitors in **series**

$$Q = Q_1 = Q_2 = Q_3$$
$$V = V_1 + V_2 + V_3$$

charge on one plate

$$C_{total} = \frac{Q}{V}$$

$$V_1 = \frac{Q}{C_1}; V_2 = \frac{Q}{C_2}; V_3 = \frac{Q}{C_3}$$

$$V = Q\left(\frac{1}{C_1} + \frac{1}{C_2} + \frac{1}{C_3}\right)$$

$$\frac{1}{C_{total}} = \frac{1}{C_1} + \frac{1}{C_2} + \frac{1}{C_3}$$

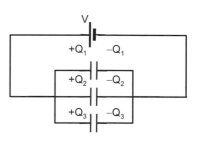

Capacitors in **parallel**

$$+Q = Q_1 + Q_2 + Q_3$$
$$V = V_1 = V_2 = V_3$$

$$C_{total} = \frac{Q}{V}$$

$$V_1 = \frac{Q_1}{C_1}; V_2 = \frac{Q_2}{C_2}; V_3 = \frac{Q_3}{C_3}$$

$$Q = V_1 C_1 + V_2 C_2 + V_3 C_3$$
$$= V(C_1 + C_2 + C_3)$$

$$\frac{Q}{V} = C_1 + C_2 + C_3$$
$$= C_{total}$$

3.1.17 Kirchhoff's laws

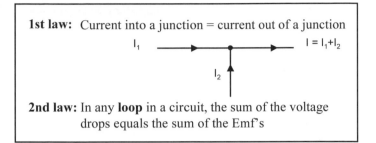

1st law: Current into a junction = current out of a junction

$I = I_1 + I_2$

2nd law: In any **loop** in a circuit, the sum of the voltage drops equals the sum of the Emf's

Example:

In the circuit shown, calculate R_1, and the current through the section A-B

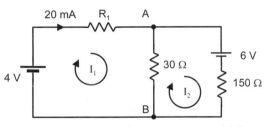

Solution:

1. Divide the circuit up into **current loops** and draw an arrow which indicates the direction of current assigned to each loop (the direction you choose need not be the correct one. If you guess wrongly, then the current will simply come out negative in the calculations).

2. Consider each loop separately:

Current going the right way through a voltage source is positive.

Current going the wrong way is negative.

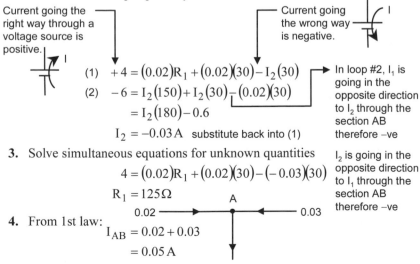

(1) $+4 = (0.02)R_1 + (0.02)(30) - I_2(30)$

(2) $-6 = I_2(150) + I_2(30) - (0.02)(30)$

$= I_2(180) - 0.6$

$I_2 = -0.03\,A$ substitute back into (1)

In loop #2, I_1 is going in the opposite direction to I_2 through the section AB therefore $-ve$

3. Solve simultaneous equations for unknown quantities

$4 = (0.02)R_1 + (0.02)(30) - (-0.03)(30)$

$R_1 = 125\,\Omega$

I_2 is going in the opposite direction to I_1 through the section AB therefore $-ve$

4. From 1st law:
$I_{AB} = 0.02 + 0.03$

$= 0.05\,A$

3.1.18 Examples

1. A power station supplies 11kV at 1 amp along a transmission line of resistance R. Calculate the power transmitted and the power lost to heat in the transmission line. Explain why it is efficient to transmit power at high voltage and low current.

Solution:

Power transmitted is $VI = 11$ kW. Some of this electrical energy is dissipated as heat within the line at a rate of $P = \Delta VI$ where ΔV is the voltage *drop* from one end of the line to the other (not the 11 kV which is the voltage drop from the line to earth). The power dissipated can also be calculated from $P = I^2R$ where I is the current in the line and R is its resistance Doubling the voltage and halving the current would reduce the power dissipated in the line by a factor of four and increase that available to the consumer by the same amount. Thus, it is better to transmit at high voltage and a low current.

2. Explain what happens when the distance between a parallel plate capacitor across which is maintained a steady voltage V is increased.

Solution:

Increasing the distance d results in a decrease in field strength E since *in the example shown here,* V is a constant and $V = Ed$. This decrease in E must result in a decrease in the accumulated charge Q on each plate since $E = \dfrac{Q}{\varepsilon_0 A}$

What happens to the excess charge? If d is increased, and the external voltage V is maintained across the capacitor, then there is a momentary increase in the opposing voltage across the capacitor. That is, the excess charge flows back into the voltage source.

A decrease in accumulated charge for a given applied voltage means a decrease in capacitance C in accordance with: $C = \varepsilon_0 \dfrac{A}{d}$

3.2 Magnetism

Summary

$$F = qv \times B$$
Force on a moving charge in magnetic field

$$B = \frac{\Phi}{A}$$
Magnetic induction

$$R = \frac{mv}{Bq}$$
Radius of curvature of moving charge in magnetic field

$$\overline{F} = I\overline{l} \times \overline{B}$$
Force on a current carrying conductor

$$\text{Torque} = IAB\sin\theta$$
Torque from electric motor

$$\overline{B} = k'\frac{q\overline{v} \times \hat{r}}{r^2}$$
Magnetic field - moving charge

$$\overline{B} = k'\int\frac{Id\overline{l} \times \hat{r}}{r^2}$$
Biot-Savart law

$$B = \frac{\mu_o I}{2\pi r}$$
Magnetic field - long straight wire

$$F = \frac{\mu_o II'l}{2\pi r}$$
Magnetic field - two parallel wires

$$B = \frac{\mu_o NI}{2a}$$
Magnetic field - flat circular coil

$$\oint B.dl = \mu_o I$$
Ampere's law

$$B = \mu_o I\frac{N}{l}$$
Solenoid

$$B = \mu_o H$$
Magnetic field intensity

$$B = B_o + \mu_o M$$
Magnetisation

$$M = \chi H$$
Susceptibility

3.2.1 Magnetic field

Over and above any electrostatic force

A **magnetic field** is said to exist at a point if a force is exerted on a moving charge at that point. The force (N) acting on a **moving charge** (C) is perpendicular to both the direction of the field and the velocity (m/s) of the charge.

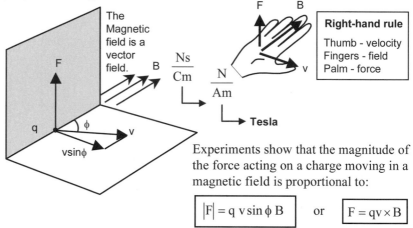

The Magnetic field is a vector field.

$B \quad \dfrac{Ns}{Cm} \quad \dfrac{N}{Am}$

→ **Tesla**

Right-hand rule

Thumb - velocity
Fingers - field
Palm - force

Experiments show that the magnitude of the force acting on a charge moving in a magnetic field is proportional to:

$$|F| = q \, v \sin \phi \, B \qquad \text{or} \qquad F = qv \times B$$

B is called the **magnetic induction** or the **magnetic field**.

A magnetic field may be represented by lines of **induction**. The **magnetic flux** is proportional to the total number of lines.

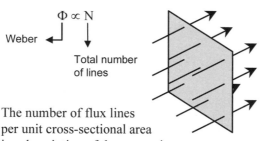

$\Phi \propto N$

Weber ←

Total number of lines

Note: not **lines of force** since, unlike the electric field, the magnetic force is perpendicular to the direction of the field.

Uniform magnetic field - lines are equally spaced.

Spacing between field lines indicates field strength.

The number of flux lines per unit cross-sectional area is a description of the magnetic field and is called the **magnetic induction** B or **magnetic flux density** (Tesla).

$$B = \dfrac{\Phi}{A}$$

1 Tesla = 1 Weber per square metre

3.2.2 Charged particle in magnetic field

1. A positively charged particle is given velocity v in a direction perpendicular to a uniform magnetic field B.

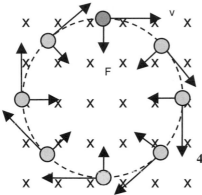

2. A force F = qvB is exerted on the particle downwards.

3. Application of force changes the direction of motion of the particle.

4. If the motion of the particle is completely within the field, then particle travels in a circle of radius R with constant **tangential speed** |v|.

   If the direction of motion is not perpendicular to the field, then the velocity component parallel to the field remains constant and particle moves in a helix

5. The force F is a **centripetal force**:

$$F = qvB$$

$$= m\frac{v^2}{R}$$

$$R = \frac{mv}{Bq}$$

Radius of path

$$\omega = \frac{qB}{m}$$

Frequency

Magnetic poles

It is a peculiar property of magnetic field lines that they always form closed loops. Electric field lines may start on an isolated positive charge and terminate on another isolated negative charge. Magnetic field lines do not start and finish on isolated **magnetic poles** even though we may draw them as starting from the North pole of a magnet and finishing on the south pole. Magnetic field lines actually pass through the magnet to join up again.

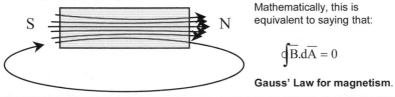

Mathematically, this is equivalent to saying that:

$$\oint B.d\overline{A} = 0$$

Gauss' Law for magnetism.

3.2.3 Force on a current-carrying conductor

Consider the movement of both positive and negative charge carriers in a conductor perpendicular to a magnetic field B

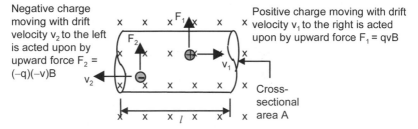

Negative charge moving with drift velocity v_2 to the left is acted upon by upward force $F_2 = (-q)(-v)B$

Positive charge moving with drift velocity v_1 to the right is acted upon by upward force $F_1 = qvB$

Cross-sectional area A

Let n_1 and n_2 be the number of positive and negative charge carriers per unit volume. The total number of charge carriers N in a length l and cross-sectional area A of the conductor is:

$$N = n_1 A l + n_2 A l$$

$$F = n_1 A l (q_1 v_1 B) + n_2 A l (q_2 v_2 B)$$
$$= (n_1 q_1 v_1 + n_2 q_2 v_2) A l B$$
$$= J A l B \qquad J \longrightarrow \text{Current density}$$
$$= I l B$$

Total force on all charge carriers both positive and negative

$$\boxed{\overline{F} = I\overline{l} \times \overline{B}}$$

Note: Here we have considered the movement of both positive and negative charge carriers within a **conductor**. If current flows due to the movement of only one type of charge, (e.g. electrons in a metal), then, from the macroscopic point of view, this is *exactly* equivalent to the equal movement of only positive charge carriers in the opposite direction.

The resultant force on the loop is zero
The resultant **torque** (or moment) is:

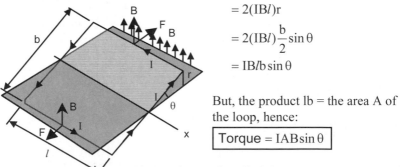

$$M_x = 2Fr$$
$$= 2(IBl)r$$
$$= 2(IBl)\frac{b}{2}\sin\theta$$
$$= IBlb\sin\theta$$

But, the product lb = the area A of the loop, hence:

$$\boxed{\text{Torque} = IAB\sin\theta}$$

The product IA is called the **magnetic moment** of the loop and is (unfortunately) given the symbol: μ.

3.2.4 Source of magnetic fields

A **magnetic field** can be created in two ways: (a) by the movement of charge carriers in a conductor (i.e. an electric current) and (b) by a changing **electric field** in an insulator (or empty space).

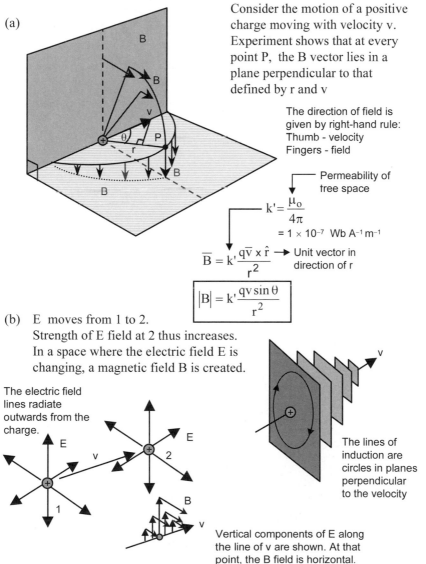

(a)

Consider the motion of a positive charge moving with velocity v. Experiment shows that at every point P, the B vector lies in a plane perpendicular to that defined by r and v

The direction of field is given by right-hand rule:
Thumb - velocity
Fingers - field

Permeability of free space

$$k' = \frac{\mu_o}{4\pi}$$
$$= 1 \times 10^{-7} \ \text{Wb A}^{-1}\text{m}^{-1}$$

$$\overline{B} = k' \frac{q\overline{v} \times \hat{r}}{r^2} \longrightarrow \text{Unit vector in direction of r}$$

$$|B| = k' \frac{qv \sin\theta}{r^2}$$

(b)   E moves from 1 to 2.
Strength of E field at 2 thus increases.
In a space where the electric field E is
changing, a magnetic field B is created.

The electric field
lines radiate
outwards from the
charge.

The lines of
induction are
circles in planes
perpendicular
to the velocity

Vertical components of E along
the line of v are shown. At that
point, the B field is horizontal.

3.2.5 Biot-Savart law

Consider an element of **conductor** length dl and area A carrying n charge carriers with drift velocity c. The total charge moving through a volume element is:

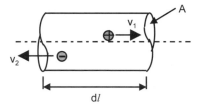

Number density of charges ⟶ ⟵ Volume

$$dQ = nq(Adl)$$

Now, $|dB| = k'\dfrac{(dQ)v\sin\theta}{r^2}$

$$= k'\dfrac{(nqAdl)v\sin\theta}{r^2}$$

$$= k'\dfrac{Idl\sin\theta}{r^2} \quad \text{since nqvA = I}$$

In vector notation:

$$d\overline{B} = k'\dfrac{Id\overline{l} \times \hat{r}}{r^2}$$

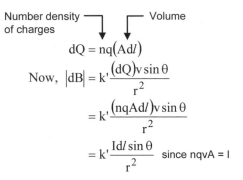

$$\boxed{\overline{B} = k'\int\dfrac{Id\overline{l}\times\hat{r}}{r^2}}$$

Add field from all elements to get total field B.

Biot-Savart Law

This law demonstrates how a steady current I in a conductor produces a magnetic field B. It is one of the two ways in which a magnetic field may be created. The other way is by the presence of a **changing electric field** in space.

3.2.6 Biot-Savart law - Applications

(1) A <u>long</u> **straight wire**

$$k' = \frac{\mu_o}{4\pi}$$

$\mu_o = 4\pi \times 10^{-7}\,Wb\,A^{-1}\,m^{-1}$

$$B = 2k'\frac{I}{r}$$

Permeability of free space

$$\boxed{B = \frac{\mu_o I}{\pi 2r}}$$

Perpendicular distance from wire.

The contribution from each moving charge comprising the current I results in a steady magnetic field in space which encircles the conductor and diminishes as 1/r.

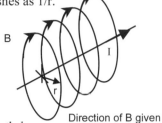

Direction of B given by right-hand rule.

(2) Force between **two parallel wires** of length l

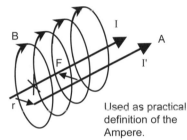

Used as practical definition of the Ampere.

Each conductor experiences a force
Consider the force on conductor A

$$F = BI'l$$

but $\quad B = \frac{\mu_o I}{2\pi r}$

$$\boxed{F = \frac{\mu_o II'l}{2\pi r}}$$

Right-hand rule shows that wires carrying currents in same direction attract and in opposite directions, repel.

(3) **Flat circular coil** (N coils)

$$B = \frac{\mu_o NIa^2}{2(x^2 + a^2)^{3/2}}$$

$$\boxed{B = \frac{\mu_o NI}{2a}} \quad @ \; x = 0$$

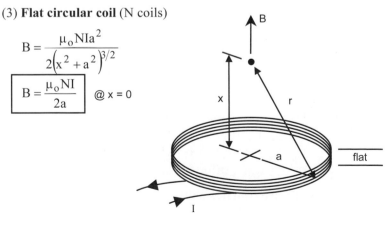

3.2.7 Ampere's law

Ampere's law is an alternative statement to **Biot-Savart** law:

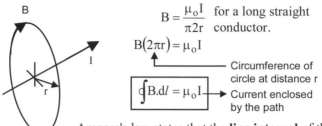

$$B = \frac{\mu_0 I}{\pi 2r} \quad \text{for a long straight conductor.}$$

$$B(2\pi r) = \mu_0 I$$

$$\oint B.dl = \mu_0 I$$

— Circumference of circle at distance r

→ Current enclosed by the path

Ampere's law states that the **line integral** of the magnetic induction field around any closed path is proportional to the net current through any area enclosed by the path.

The line integral is a special type of integral in which only the component of the function being integrated acting in the same direction of the direction of the path of integration is summed.	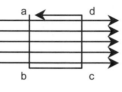 The dot product B.dl is zero for the path segments ab and cd since those path segments are perpendicular to B. The only contribution to the line integral are the segments ad and bc.

The magnetic field of a current-carrying conductor may be concentrated by winding the conductor around a tube to form a **solenoid**.

Single conductor

Stranded conductor

Single loop of a **stranded conductor**: i.e. a solenoid

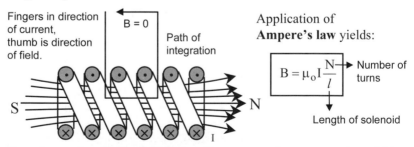

Fingers in direction of current, thumb is direction of field.

B = 0

Path of integration

Application of **Ampere's law** yields:

$$B = \mu_0 I \frac{N}{l}$$

→ Number of turns

Length of solenoid

For a *long* solenoid, the field B is uniform across the cross-section within the solenoid. The field decreases slightly near the ends of the solenoid (as shown by the wider spacing between the flux lines in the above diagram.

3.2.8 Magnetic moment

Experiments show that:

- a charge moving perpendicular to magnetic field lines experiences a magnetic force.

$$|F| = qvB\sin\phi$$

- a magnetic field is produced by a moving charge.

$$|B| = k'\frac{qv\sin\theta}{r^2}$$

These two phenomena are a consequence of the natural tendency of magnetic fields to align themselves. Consider the magnetic field created by the moving charges in a solenoid coil:

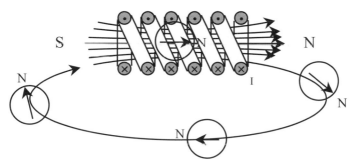

The ends of such a magnetised coil are commonly labelled **North** and **South** to indicate the direction of the field. These labels have come about since it is the direction in which the solenoid would tend to align itself with the magnetic field of the earth if it were free to rotate.

A compass needle itself consists of a North and South pole, it being a small magnet free to rotate on a spindle. A compass needle placed in the vicinity of the coil would tend to align itself with the field surrounding the coil. Now, outside the coil, the alignment results in the familiar observation that like poles repel and unlike poles attract. Inside the coil, the compass needle is still aligned with the field but we can no longer say that like poles repel and unlike poles attract. Rather, it is more scientifically appropriate to say that *when two magnetic fields interact, they tend to align themselves.* This tendency to alignment exerts a mutual torque (or moment) between the bodies producing the fields. This is called the **magnetic moment**.

3.2.9 Magnetic force

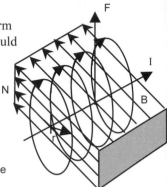

Consider a **current carrying wire** in a uniform magnetic field. We know that such a wire would experience a force upwards as shown in the diagram. How does this force come about?

The total field (given by the vector sum of the **flux lines**) is stronger at the bottom and weaker at the top.

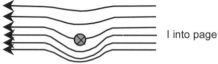

I into page

If we think of flux lines being under tension then it is easy to imagine the increased tension in the bottom lines of flux tending to push upwards on the wire.

Alternately, imagine the current-carrying wire is attached to a pivot point by a rigid beam. The **magnetic moment** arising from the tendency of the two magnetic fields to align themselves acts around this pivot point. The wire itself is thus subjected to an upwards force as shown. The force is a maximum where the moment arm r is the greatest.

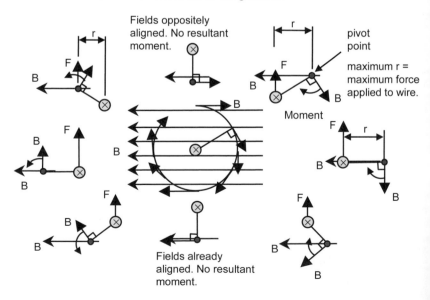

3.2.10 Permeability

When matter is placed in the region surrounding a current-carrying conductor, the magnetic field is different to that which exists when the conductor is in a vacuum due to the **magnetisation** of the material.

1. Uniform field

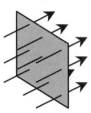

2. Presence of material may concentrate field

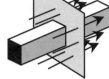

This behaviour arises due to the interaction of the externally applied field and the internal field generated by the orbiting (and also spinning) electrons within the material

Now, a magnetic field can be described by either of two vectors:

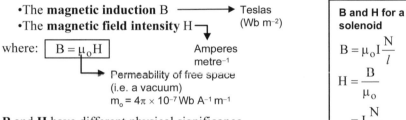

- The **magnetic induction** B $\longrightarrow$ Teslas (Wb m^{-2})
- The **magnetic field intensity** H

where: $\boxed{B = \mu_o H}$ Amperes metre^{-1}

$\hookrightarrow$ Permeability of free space (i.e. a vacuum)
$m_o = 4\pi \times 10^{-7}$ Wb A^{-1} m^{-1}

B and H for a solenoid

$$B = \mu_o I \frac{N}{l}$$

$$H = \frac{B}{\mu_o}$$

$$= I \frac{N}{l}$$

B and **H** have different physical significance.

The magnetic induction **B** is a field vector which determines the magnetic force acting on a moving charged particle. $\Big\}$ a magnetic "effect"

The magnetic field intensity **H** is also a field vector, but describes the magnetic field *generated* by a moving charged particle. $\Big\}$ a magnetic "cause"

The constant μ_o shows that the resulting magnetic induction B produced by an external source of magnetic field (such as a permanent magnet or moving charge) depends upon the "ease" with which the surrounding material permits the creation of magnetic field lines.

μ_o is called the **permeability** of the material. Those materials which tend to facilitate the setting-up of magnetic field lines have a higher permeability than others.

3.2.11 Magnetic materials

The behaviour of materials when placed in a magnetic field is a consequence of three types of behaviour:

• **paramagnetism** ⟶ Internal magnetic fields line up and
• **diamagnetism** ⟶   reinforce the external field.
• **ferromagnetism** ↓ Atoms whose shells are not completely filled.

Internal fields add up in such a way so as to oppose the external field.

Most prominent in atoms whose shells are completely filled.

Internal fields are strongly interacting with each other lining up into **magnetic domains**, even when there is no external field present.

Internally generated magnetic fields of the material add to the external field, i.e. that which would exist in the absence of any material through which magnetic flux passes and is described by the quantity $\mu_o H$. The additional field produced is proportional to the total **magnetic moment** per unit volume in the material. In an atom, the magnetic moment arises when atoms (whose orbiting electrons generate a magnetic field) attempt to line up with the externally applied field.

The total magnetic moment per unit volume is called the **magnetisation M**:

$$M = \frac{\mu_{total}}{V}$$

The additional field due to magnetisation is given by $\mu_o M$. Thus, when the conductor is completely surrounded by material, the magnetic field B in the material is:

$$B = B_o + \mu_o M$$

Total field within the material when current flows in conductor.

External field generated from current $B_o = \mu_o H$

Additional field due to magnetisation of material.

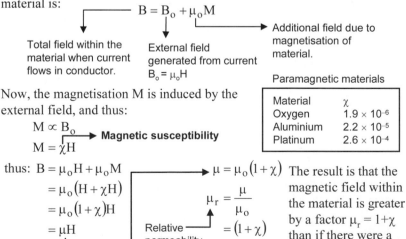

Paramagnetic materials

Material	χ
Oxygen	1.9×10^{-6}
Aluminium	2.2×10^{-5}
Platinum	2.6×10^{-4}

Now, the magnetisation M is induced by the external field, and thus:

$$M \propto B_o$$

$$M = \chi H$$

⟶ **Magnetic susceptibility**

thus: $B = \mu_o H + \mu_o M$
$= \mu_o (H + \chi H)$
$= \mu_o (1 + \chi)H$
$= \mu H$

$\mu_r = \dfrac{\mu}{\mu_o}$

Relative permeability $= (1 + \chi)$

$\mu = \mu_o(1 + \chi)$ The result is that the magnetic field within the material is greater by a factor $\mu_r = 1 + \chi$ than if there were a vacuum present.

3.2.12 Ferromagnetism

Ferromagnetic materials have permeabilities much larger than that of free space.

$\mu_r = 1000\text{-}10000$ This large relative permeability arises due to an appreciable net magnetic moment due to partially-filled 3rd electron shells.

The net magnetic moments interact within the material even when there is no external field present.

These interactions cause neighbouring moments to align themselves parallel to each other in regions called **magnetic domains**.

With no external field, the orientations of the domains are random. When a field H is applied, the domains attempt to align themselves with the field. Those domains already in alignment with the field tend to grow at the expense of others, not aligned, which shrink. As H is increased, a point is reached where all domains are aligned with the field.

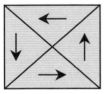
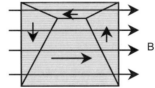

Iron is the most significant magnetic material hence the term **ferromagnetic** but the term also refers to other elements such as Nickel and Cobalt.

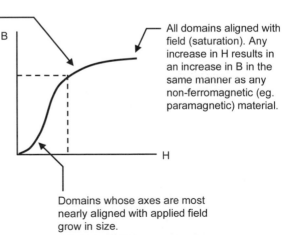

The permeability of a ferromagnetic material varies as the magnetic field intensity H is varied. For a given value of H, B depends on μ because H is great enough to force domains to align themselves with H rather than their preferred crystalline orientation, or, m depends on B Because the permeability depends upon what fraction of the magnetic domains have aligned their moments with the magnetic field.

All domains aligned with field (saturation). Any increase in H results in an increase in B in the same manner as any non-ferromagnetic (eg. paramagnetic) material.

Domains whose axes are most nearly aligned with applied field grow in size.

A BH curve shows how flux density of an <u>unmagnetised</u> sample of ferromagnetic material rises as the value of the field H is increased.

3.2.13 Example

1. A solenoid has length 20 cm, diameter 1 cm and 1000 turns. What current must flow in the coil to develop a total flux of 1×10^{-6} Wb at the centre of the solenoid?

Solution:

$$B = \frac{\phi}{A}$$

$$= \mu_o I \frac{N}{l}$$

$$I = \frac{\phi}{A} \frac{l}{\mu_o N}$$

$$= \frac{1 \times 10^{-6} (0.2)}{\pi (0.005^2)(4\pi \times 10^{-7})(1000)}$$

$$= 2.206 \, A$$

Hysteresis

Most magnetic materials will show some residual magnetism when the field is returned to zero:

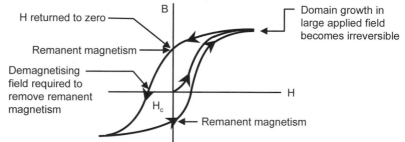

H returned to zero

Remanent magnetism

Demagnetising field required to remove remanent magnetism

H_c

B

Domain growth in large applied field becomes irreversible

H

Remanent magnetism

When the magnetic field intensity H arises from an alternating current, the flux density B in a ferromagnetic material tends to lag behind the magnetic field intensity H which creates it. This is called **hysteresis**.

The **hysteresis loop** represents an energy loss as the field has to be reversed to negate the residual magnetism arising from the irreversible alignment of domains at high values of H.

Only **ferromagnetic materials** have **residual magnetism** and thus show hysteresis effect. The area within hysteresis loop is an indication of energy loss thus different materials may be compared experimentally.

3.3 Induction

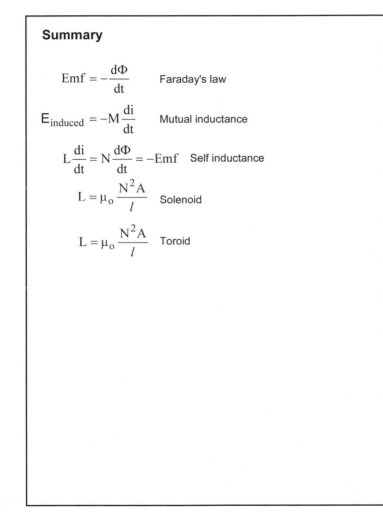

Summary

$$\text{Emf} = -\frac{d\Phi}{dt} \qquad \text{Faraday's law}$$

$$\text{E}_{\text{induced}} = -M\frac{di}{dt} \qquad \text{Mutual inductance}$$

$$L\frac{di}{dt} = N\frac{d\Phi}{dt} = -\text{Emf} \qquad \text{Self inductance}$$

$$L = \mu_o \frac{N^2 A}{l} \qquad \text{Solenoid}$$

$$L = \mu_o \frac{N^2 A}{l} \qquad \text{Toroid}$$

3.3.1 Faraday's law

Consider a loop of conducting wire in a magnetic field. One side of the conducting loop is made as a slider which can be moved.

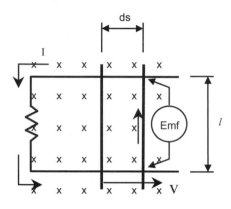

When the slider is moved to the right with velocity v, the (positive) charges in the moving wire experience a force $F = qvB$ upwards thus causing a (conventional) current to flow anti-clockwise in the loop.

Note: The slider must "cut" the magnetic field B.

When the conductor moves to the right a distance ds, the cross-sectional area swept out by the movement is: $dA = l\,ds$.

Now, the **magnetic flux** Φ is given by:

$BA = \Phi$ ← Proportional to the number of lines of magnetic induction in the B field

$d\Phi = B\,dA$ ← Change in flux when wire moves a distance ds

$= Bl\,ds$

$\dfrac{d\Phi}{dt} = Bl\dfrac{ds}{dt}$ ← Velocity v

$= vBl$

$= |Emf|$

$$\boxed{Emf = -\dfrac{d\Phi}{dt}}$$

Faraday's Law

$F = qvB$

$Fl = qvBl$

$\Delta PE = qvBl$

$\dfrac{\Delta PE}{q} = vBl$

$= Emf$

Sign conventions:
- If the current is clockwise, the Emf is positive.
- If the direction of B is away from the observer (as shown above) then the increase in Φ is positive.

The induced Emf in the circuit is equal to the **rate of change** of **magnetic flux** through it.

3.3.2 Electromagnetic induction

From **Faraday's law**, we have: $\text{Emf} = -\dfrac{d\Phi}{dt}$

An Emf is induced in a conductor whenever there is a change of **magnetic flux**. Now, a change of flux can be brought about in two ways:

* the conductor may move in the magnetic field and "cut" lines of B.
* a varying magnetic field can cause an induced Emf in a stationary conductor.

If the current I is changing, then the flux density also changes. If the current increases, more lines of flux move outwards from the conductor to encircle the conductor.

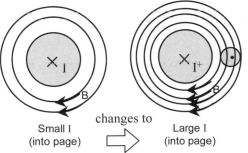

Small I
(into page)

changes to

Large I
(into page)

A second conductor placed nearby experiences this changing flux and a voltage is induced within it. If the conductor forms part of a circuit, then current flows (out of page in this example). Same effect happens if second conductor is moved to the left.

Consider the second conductor: I (out of page)

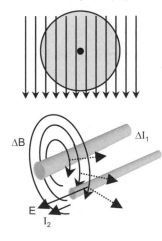

Magnetic field lines from the increasing current in the first conductor appear to move to the right, which is the same as saying that the second conductor moves towards the left. That is, charges in the second conductor move through space towards the left relative to the lines of B. By Right Hand Rule, charges experience a force out of page, i.e. current in second conductor is out of the page.

Alternately, we can say that in the space surrounding the second conductor, there is a changing magnetic field as the flux lines move outwards and become closer together. A changing magnetic field creates an electric field perpendicular to it and the direction of motion of the B flux lines. This electric field E (i.e. an induced Emf) then acts upon the stationary charges in the second conductor causing them to move (if the conductor is part of a circuit). Same effect if second conductor is moved inwards towards the first conductor. If the change in B is "linear" (dB/dt=constant) then the induced current I_2 is constant.

3.3.3 Lenz's law

1. Consider the example of induction given earlier:

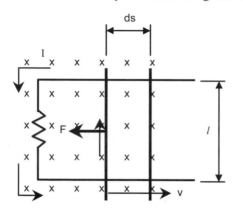

The (positive) charges in the moving wire experience a force $F = qvB$ upwards thus causing a current to flow anti-clockwise.

However, a current carrying conductor within a magnetic field experiences a magnetic force $F = Il \times B$, which by the right hand rule, is towards the left in this example.

The direction of the induced current in the moving wire is such that the direction of the resulting magnetic force is opposite in direction to its motion. The work done in maintaining the motion v is dissipated as heat in the circuit (here shown as a resistor) by the passage of current through it.

2. Now consider the case where a changing magnetic flux induces a current in a stationary conductor (which forms part of a circuit):

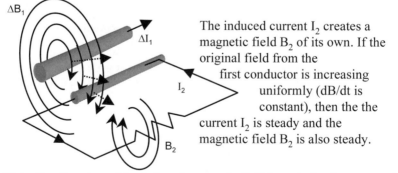

The induced current I_2 creates a magnetic field B_2 of its own. If the original field from the first conductor is increasing uniformly (dB/dt is constant), then the the current I_2 is steady and the magnetic field B_2 is also steady.

Note, the direction of the induced magnetic field is such that it opposes the change in flux through a loop in the circuit. If the field B_1 is increasing, the induced field inside the circuit loop is in the opposite direction tending to oppose the increase. If the field B_1 is decreasing, current in the second circuit flows in the opposite direction to that shown above and the field B_2 is in the same direction, tending to oppose the decrease. **Lenz's Law**

3.3.4 Mutual inductance

If a varying current flows in coil #1, a varying magnetic field is produced. This varying magnetic field can then in turn induce an Emf (and thus a current) in another coil #2.

Lower case i signifies an instantaneous value of a varying current.

$$E_{induced} = -M \frac{di}{dt}$$

A constant which depends on the nature of the coils. The **mutual inductance**.

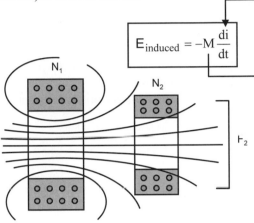

N_1

N_2

$\vdash_2$

A portion of the flux set up by current in coil #1 links with coil #2. Thus, there is a flux passing through each turn in coil #2.

The product $N_2\Phi_2$ is called the **number of flux linkages**.

The mutual inductance M is defined as the ratio of the number of flux linkages in coil #2 to the current in coil #1: i_1

$$M = \frac{N_2\Phi_2}{i_1}$$

But, if the current varies with time, then so does the number of flux linkages, (both i and Φ are functions of t) thus:

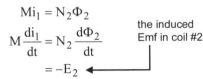

$$Mi_1 = N_2\Phi_2$$

$$M\frac{di_1}{dt} = N_2 \frac{d\Phi_2}{dt}$$

the induced Emf in coil #2

$$= -E_2$$

For a given mutual inductance M, the greater the rate of change of i, the greater the induced Emf in coil #2.

The mutual inductance can be considered as the induced Emf in coil #2 per unit rate rate of change of current in coil #1.

M = 1 volt per ampere per second

1 Henry

3.3.5 Self-inductance

As the current increases in a single conductor, lines of induction are created and move outwards as the current increases.

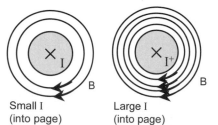

Small I
(into page)

Large I
(into page)

As the lines of induction move across the conductor material, they generate a voltage within the conductor itself.

The **self-induced voltage** is opposite to that which produces the original current.

Consider left side of conductor:

Magnetic field lines move outwards to the left which is the same as the charges in the conductor moving to the right.

By right hand rule, force on charges causes induced current to flow out of page which opposes original current flowing into page.

Consider right side of conductor:

Magnetic field lines move outwards to the right which is the same as the charges in the conductor moving to the left.

By right hand rule, force on charges causes induced current to flow out of page which opposes original current flowing into page.

The self-induced current represents a self-induced voltage opposite to that applied to the wire. It is quite small for a single conductor, but can be made significant when the conductor is formed into a coil, or **solenoid** since as the current builds up in first loop, the changing magnetic field induces a voltage in second loop which tends to oppose the applied voltage and so on.

$$L = \frac{N\Phi}{i}$$

Inductance   ← Instantaneous current

If the current I changes with time, then so does Φ and thus:

$$L = \frac{N \dfrac{d\Phi}{dt}}{\dfrac{di}{dt}}$$

$$L \frac{di}{dt} = N \frac{d\Phi}{dt}$$

$$= -\text{Emf}$$

A flux Φ passes through each turn. The number of flux linkages for all coils is $N\Phi$. The number of linkages per unit of current is called the **self inductance**.

The self inductance is the self-induced Emf (or **back Emf**) per unit rate of change of current. The direction of the self-induced Emf is found by **Lenz's Law**.

3.3.6 Solenoids and toroids

Inductance of a **solenoid**:

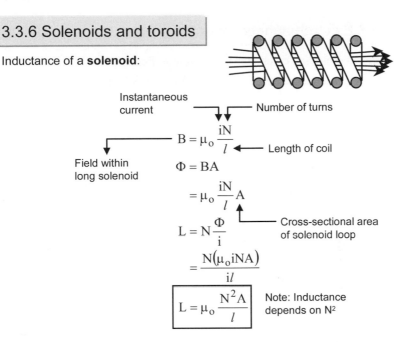

Instantaneous current — Number of turns

$$B = \mu_o \frac{iN}{l}$$ ← Length of coil

Field within long solenoid

$$\Phi = BA$$

$$= \mu_o \frac{iN}{l} A$$ — Cross-sectional area of solenoid loop

$$L = N\frac{\Phi}{i}$$

$$= \frac{N(\mu_o iNA)}{il}$$

$$\boxed{L = \mu_o \frac{N^2 A}{l}}$$ Note: Inductance depends on N^2

Inductance of a **toroid**:

The case of a toroid is somewhat simpler than that of the solenoid because the magnetic field is confined wholly within the toroid.

→ and is uniform if the radius of the toroid is large with respect to the radius of the turns.

$$\boxed{L = \mu_o \frac{N^2 A}{l}}$$

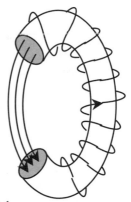

The length l is the circumference of the toroid and the area A is the cross-sectional area of the loops.

The use of μ_o here signifies the inductance of a toroid with an air (or strictly speaking, vacuum) core. The inductance of a toroid may be significantly increased when the coil is wound on a material with a high **permeability**.

3.3.7 Electromagnetic induction   (Magnetic field approach)

Consider a moving conductor in a magnetic field:

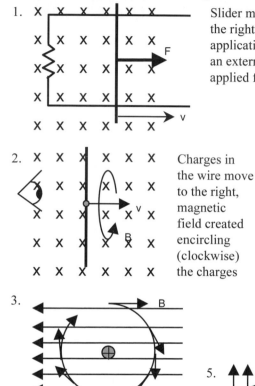

1.

Slider moves to the right by the application of an externally applied force F.

2.

Charges in the wire move to the right, magnetic field created encircling (clockwise) the charges

3.

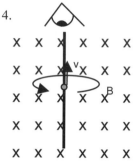

Looking in from the left, charge experiences an upwards force F = qvB due to the tendency of the magnetic fields to align themselves.

4.

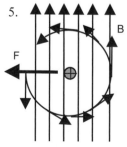

Charges move upwards through wire (electric current) under the influence of F = qvB. Another magnetic field encircles the moving charges.

5.

Looking down from top, charges experience a sideways force F = qvB due to the tendency of the magnetic fields to align themselves. This force F acts against the original force F which is moving the slider (Lenz's law).

3.3.8 Examples

1. A coil has 1000 turns, length 300 mm and diameter 40 mm wound on a cardboard former. Determine the inductance of the coil and the energy stored within it when carrying a current of 1.5 A.

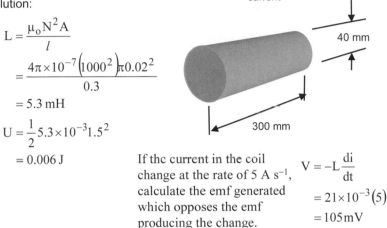

Final steady-state current

40 mm

300 mm

Solution:

$$L = \frac{\mu_0 N^2 A}{l}$$

$$= \frac{4\pi \times 10^{-7} \left(1000^2\right) \pi 0.02^2}{0.3}$$

$$= 5.3 \text{ mH}$$

$$U = \frac{1}{2} 5.3 \times 10^{-3} 1.5^2$$

$$= 0.006 \text{ J}$$

If the current in the coil change at the rate of 5 A s⁻¹, calculate the emf generated which opposes the emf producing the change.

$$V = -L \frac{di}{dt}$$

$$= 21 \times 10^{-3} (5)$$

$$= 105 \text{ mV}$$

2. Consider a rotating, conducting (non magnetic) disk through which passes a magnetic field. Will there be an induced current in the disk?

Solution:

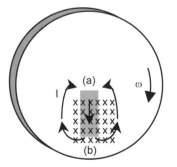

1. The shaded area moves through the field and a current is induced in that portion of the disk directed downwards from (a) to (b)

2. Neighbouring portions of the disk are not in the field but provide conducting paths for the current to flow back to (a)

3. The circulation of charge in this manner are called **eddy currents**.

The downwards current from (a) to (b) experiences a sideways force F = IlB which opposes the rotation of the disk (**Lenz's law**). Hence eddy currents represent an energy loss (due to I²R heating) which may or may not be desirable.

3.4 Magnetic circuits

Summary

$F_m = NI$ Magnetomotive force

$R_m = \dfrac{F_m}{\Phi}$ Reluctance

$\mu = \dfrac{l}{R_m A}$ Permeability

$H = \dfrac{F_m}{l}$ Magnetic field intensity

3.4.1 Magnetomotive force

The magnetic field of a current-carrying conductor may be concentrated by winding the conductor around a tube to form a **solenoid**.

Application of Ampere's law yields:

$$B = \mu_o I \frac{N}{l}$$

N ◄——— Number of turns
l ◄——— Length of solenoid

For a *long* solenoid, the field B is uniform across the cross-section within the solenoid.

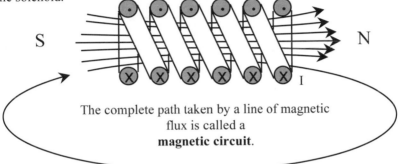

S   N

I

The complete path taken by a line of magnetic flux is called a **magnetic circuit**.

The solenoid concentrates the magnetic field within the space enclosed by its windings. Although shown only for one line in the above figure, all lines of magnetic induction form complete loops, but the spacing of the lines outside the solenoid is very large (small value of B compared to inside coil).

Lines of magnetic flux only appear when there is current in the coil. Magnetic flux lines, can be imagined to be established through the action of a **magnetomotive force**. The magnitude of the magnetomotive force is evidently dependent on the current in the solenoid and the number of turns.

$$\boxed{F_m = NI}$$

Magnetomotive force (mmf)
units: ampere-turns

This holds even for a single coil of wire. Thus, the length term *l* is a modifying effect of the solenoid and is not included in the definition of the magnetomotive force

3.4.2 Reluctance

Lines of magnetic induction, or just **magnetic flux**, are established by a **magnetomotive force** (mmf).

$$F_m = NI$$

Lines of induction form complete loops which we may regard as a **magnetic circuit**. For a given magnetomotive force, the strength of the resulting magnetic induction field B depends upon the **permeability** μ of the region (either space or matter) where the field is created.

The term **reluctance** R_m is given to the resistance encountered by a magnetomotive force in creating a *particular* magnetic circuit.

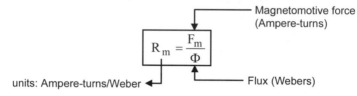

Magnetomotive force (Ampere-turns)

$$R_m = \frac{F_m}{\Phi}$$

units: Ampere-turns/Weber ← Flux (Webers)

Since R_m is a property of a particular magnetic circuit, and μ, the permeability, is a property of the region where the magnetic circuit exists, it is not surprising to find a link between these two quantities:

$$\mu = \frac{l}{R_m A}$$

Length of the magnetic circuit
Cross-sectional area of the magnetic circuit

Now, since $B = \mu H$ and $B = \Phi/A$, then:

$$\frac{\Phi}{A} = \frac{l}{R_m A} H$$

$$\frac{\Phi R_m A}{A l} = H$$

$$H = \frac{F_m}{l}$$

For **ferromagnetic materials**, the permeability μ of the material depends upon the percentage of domains which are aligned with the applied field H. That is, the reluctance is a function of the mmf since μ can no longer be considered independent of the region but depends on F_m.

This is an example of a nonlinear magnetic circuit.

$$R_m = \frac{l}{\mu A}$$

3.4.3 Magnetic circuits

The operation of electrical machinery depends upon the associated magnetic fields.

- the path of magnetic field lines can be controlled by the use of highly permeable materials (e.g. ferromagnetic materials) to construct the desired magnetic circuits appropriate to the application of the machinery involved.
- the magnetomotive force required to develop a desired magnetic field strength can be determined from: $R_m = \dfrac{F_m}{\Phi}$ sometimes called the **magnetic Ohm's law**.

Consider an **air-filled solenoid**:

Lines of induction are concentrated inside the coil (small cross-section) but are widely spaced outside the coil (large cross section).

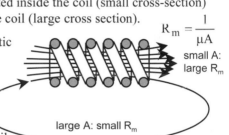

$$R_m = \frac{l}{\mu A}$$

small A: large R_m

Now, the **reluctance** of a magnetic circuit is inversely proportional to the cross-sectional area, and also, for an air-filled coil, the permeability of the air inside the coil is the same as the air outside. Thus, for an air-filled coil,

large A: small R_m

the total reluctance for the complete magnetic circuit is approximately equal to that given by just the interior portion of the coil itself since the interior area A is much smaller than the exterior A outside the coil.

The **magnetomotive force** required to establish a required magnetic flux in an air-filled coil can thus be calculated from: $F_m = R_m \Phi$

Consider an **iron-filled solenoid**:

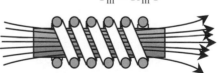

For an iron-core coil, the very high permeability of the core allows a far greater flux within the core for the same mmf even though the interior area is relatively small compared to the exterior air part of the circuit. The permeability of the core is so high that the **return path** through the surrounding air now contributes the most to the reluctance of the magnetic circuit. The situation where the return path is not specifically directed is difficult to analyse. In practice, the return path for the circuit is directed in some way.

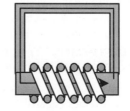

3.4.4 Magnetic circuits

The magnetic circuit shown to the right consists of two sections in **series** since the magnetic lines of induction pass through one section and then the other.

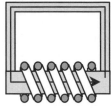

FLUX IS COMMON TO ALL PARTS OF SERIES CIRCUIT

The total **reluctance** of the circuit is the sum of the individual reluctances.

$$R_{mT} = R_{m1} + R_{m2} + R_{m3} \cdots$$

Different parts of the magnetic circuit may have different cross-sections and lengths.

$$R_m = \frac{l}{\mu A}$$

Since in a series circuit, the same flux Φ goes through all sections, and since $F_m = \Phi R_m$, then, the total mmf = sum of each mmf for each section of circuit.

$$F_{mT} = F_{m1} + F_{m2} + F_{m3} \cdots$$

Also, the total mmf = Hl, thus:

$$Hl = \Sigma F_m$$

Total length of circuit

If the material of a particular section is ferromagnetic, then a value for B must be calculated and a BH curve must be consulted to determine a value for m.

In some electric machines, it is necessary to have an **air gap** in the magnetic circuit to permit the mechanical rotation or movement of various parts etc. An air gap will increase the reluctance of a magnetic circuit. The air gap may be in fact a gap within an otherwise low reluctance circuit made from brass, paper or some other non-magnetic material.

If pole pieces have a cross-sectional area given by (m × n), then the effective cross sectional area = (m + 1)(n + 1)

Empirical correction to allow for the effects of fringing.

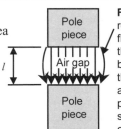

Fringing results in the flux density in the air gap being less than that within the adjacent pole pieces due to spreading out of the flux.

In a **parallel magnetic circuit**, the magnetic flux branches so that some lines go through one section and some through the other.

MMF IS COMMON TO ALL PARTS OF PARALLEL CIRCUIT

Easy solutions if structure is symmetric since flux divides equally.

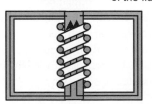

3.4.5 Example

1. Calculate the flux in the magnetic circuit shown below if the coil has
   150 turns, carries a current of 2 A and is wrapped around a magnetic
   core of cross-sectional area 0.003 m². Also calculate the reluctance of
   the magnetic circuit.

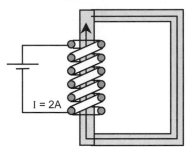

- Magnetic path
  length = 1m
- Cross-sectional
  area = 0.003 m²

I = 2A

Magnetic material	
B (T)	H (A t m⁻¹)
0.12	100
0.54	300
0.94	700
12.15	1600

Solution:

$F_m = NI$ Magnetomotive force

$\quad = 150(2)$

$\quad = 300 \, At$

$\quad = Hl$

$H = \dfrac{300}{1}$

$\quad = 300 \, At \, m^{-1}$

$B = 0.54T$ Magnetic field

$\quad = \dfrac{\Phi}{A}$

$\Phi = 0.54(0.003)$

$\quad = 0.00162 \, Wb$

$R_m = \dfrac{F_m}{\Phi}$ Reluctance

$\quad = \dfrac{300}{0.00162}$

$\quad = 185.2 \times 10^3 \, At \, Wb^{-1}$

3.5 R-C & R-L Circuits

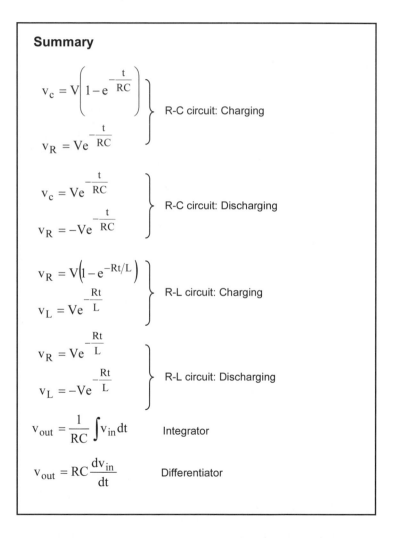

Summary

$$v_c = V\left(1 - e^{-\frac{t}{RC}}\right)$$

$$v_R = Ve^{-\frac{t}{RC}}$$

R-C circuit: Charging

$$v_c = Ve^{-\frac{t}{RC}}$$

$$v_R = -Ve^{-\frac{t}{RC}}$$

R-C circuit: Discharging

$$v_R = V\left(1 - e^{-Rt/L}\right)$$

$$v_L = Ve^{-\frac{Rt}{L}}$$

R-L circuit: Charging

$$v_R = Ve^{-\frac{Rt}{L}}$$

$$v_L = -Ve^{-\frac{Rt}{L}}$$

R-L circuit: Discharging

$$v_{out} = \frac{1}{RC}\int v_{in}\, dt$$

Integrator

$$v_{out} = RC\frac{dv_{in}}{dt}$$

Differentiator

3.5.1 R-C circuit analysis

Consider a series circuit containing a resistor R and capacitor C. If connection (a) is made, the initial potential difference across the capacitor is zero, the entire battery voltage appears across the resistor. At this instant, the current in the circuit is:

$$I_0 = \frac{V}{R}$$

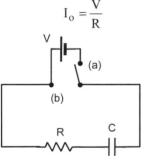

(a)

(b)

R   C

As the capacitor charges, the voltage across it increases and the voltage across the resistor correspondingly decreases (and so does the current in the circuit). After a "long" time, the capacitor is fully charged and the entire battery voltage appears across the capacitor. The current thus drops to zero and there is no potential drop across the resistor (since there is no current).

When connection (b) is made, the capacitor discharges through the resistor. The current is initially high and then over time drops to zero.

$$v_R = iR; \quad v_C = \frac{q}{C}$$

$$V = v_R + v_C \quad \text{Small letters}$$
$$= iR + \frac{q}{C} \quad \begin{array}{l}\text{signify}\\\text{instantaneous}\\\text{values}\end{array}$$

$$i = \frac{V}{R} - \frac{q}{RC} \quad \text{but } i = \frac{dq}{dt}$$

$$\frac{dq}{dt} = \frac{V}{R} - \frac{q}{RC}$$

$$\frac{dq}{VC - q} = \frac{dt}{RC} \quad \begin{array}{l}\text{1st order differential}\\\text{equation - integrate}\\\text{both sides}\end{array}$$

$$-\ln(VC - q) = \frac{t}{RC} + \text{constant}$$

$$-\ln(VC - q) = \frac{t}{RC} - \ln VC \quad \begin{array}{l}\text{When t = 0,}\\\text{q =0.}\end{array}$$

$$\ln(VC - q) - \ln VC = -\frac{t}{RC} \qquad VC = Q_f$$

$$\ln\frac{VC - q}{VC} = -\frac{t}{RC} \quad \begin{array}{l}\text{At fully charged}\\\text{capacitor has}\\\text{charge } Q_f \text{ and}\\\text{potential}\\\text{difference = V}\end{array}$$

$$1 - \frac{q}{VC} = e^{-\frac{t}{RC}}$$

$$q = VC\left(1 - e^{-\frac{t}{RC}}\right)$$

$$\boxed{q = Q_f\left(1 - e^{-\frac{t}{RC}}\right)}$$

Both current in circuit and charge on capacitor are an exponential functions of time

$$\frac{dq}{dt} = \frac{V}{R} e^{-\frac{t}{RC}} \quad \begin{array}{l}I_0 \text{ is the initial}\\\text{current in the}\\\text{circuit}\end{array}$$

$$= I_0 e^{-\frac{t}{RC}}$$

$$= i$$

$$\boxed{i = I_0 e^{-\frac{t}{RC}}}$$

3.5.2 Time constant and half-life

Charging

$$q = Q_f\left(1 - e^{-\frac{t}{RC}}\right) \quad \text{but} \quad q = \frac{v_c}{C}$$

$$\text{and} \quad \frac{Q_f}{C} = V$$

$$v_c = V\left(1 - e^{-\frac{t}{RC}}\right) \quad\longleftarrow\quad \text{thus}$$

$$v_R = Ve^{-\frac{t}{RC}} \quad \text{since:} \quad V = v_R + v_c$$

$$v_R = V - v_c$$

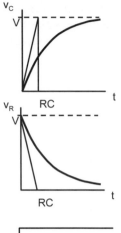

Discharging

$$0 = v_R + v_c$$

$$= iR + \frac{q}{C}$$

$$= R\frac{dq}{dt} + \frac{q}{C}$$

⇩ integrating

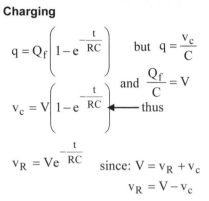

$$q = Qe^{-\frac{t}{RC}}$$

$$v_c = Ve^{-\frac{t}{RC}}$$

$$v_R = -Ve^{-\frac{t}{RC}}$$

The product RC is called the **time constant** of the circuit. It is the time in which the current (and thus v_R) would decrease to zero if it continued to decrease at its initial rate.

$$\frac{dv_c}{dt} = \left(\frac{V}{RC}\right)e^{-\frac{t}{RC}}$$

$$@\,t = 0$$

$$\frac{dv_c}{dt} = \frac{V}{RC}$$

initial slope

The **half-life** of the circuit is the time for the current to decrease to half of its initial value.

$$i = I_o e^{-\frac{t}{RC}}$$

$$\frac{I_o}{2} = I_o e^{-\frac{t_h}{RC}}$$

$$\frac{1}{2} = e^{-\frac{t_h}{RC}}$$

$$\ln\frac{1}{2} = -\frac{t_h}{RC}$$

$$\ln 1 - \ln 2 = -\frac{t_h}{RC}$$

$$t_h = RC\ln 2$$

$$= 0.693RC$$

3.5.3 R-C low pass filter

As capacitor charges, voltage across it increases. When capacitor is fully charged, all of V_{in} appears across it and no voltage drop across resistor. As time constant becomes smaller than the period T of the input pulse, the capacitor has time to charge and discharge fully.

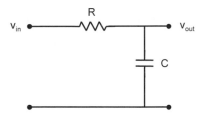

For **low pass filter**, we want RC $\ll$ T to pass through low frequencies. i.e. small time constant

Consider a square wave input signal. For a small time constant or a low frequency input signal, the capacitor charges up quickly and so the output signal looks like the input signal. This circuit acts like a low pass filter, high frequencies are shorted to ground through the capacitor.

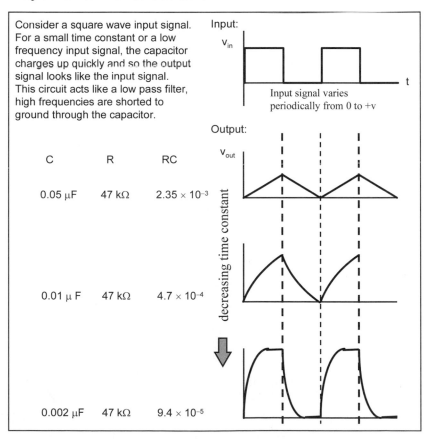

C	R	RC
0.05 µF	47 kΩ	2.35 × 10⁻³
0.01 µF	47 kΩ	4.7 × 10⁻⁴
0.002 µF	47 kΩ	9.4 × 10⁻⁵

3.5.4 R-C high pass filter

As capacitor charges, voltage across it increases and voltage across resistor decreases. As time constant becomes smaller than the period T of the input pulse, the capacitor has time to charge and discharge fully and voltage across the resistor decreases to zero.

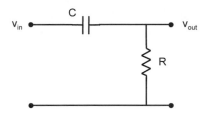

For **high pass filter**, want RC >> T to pass through high frequencies. i.e. large time constant

For a large time constant the capacitor takes a long time to charge up and so the output signal looks like the input signal. This circuit blocks low frequency input signals.

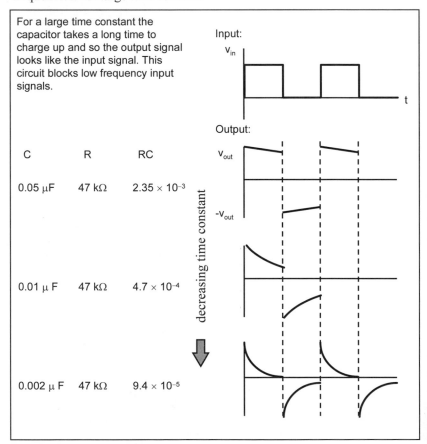

C	R	RC
0.05 μF	47 kΩ	2.35×10^{-3}
0.01 μF	47 kΩ	4.7×10^{-4}
0.002 μF	47 kΩ	9.4×10^{-5}

decreasing time constant

3.5.5 R-L circuits

Charging

$$V = v_R + v_L$$

$$= iR + L\frac{di}{dt}$$

⬇

$$i = I\left(1 - e^{-Rt/L}\right)$$

$$v_R = V\left(1 - e^{-Rt/L}\right)$$

$$v_L = Ve^{-\frac{Rt}{L}}$$

The quantity L/R is the **time constant** of the circuit and I is the final current.

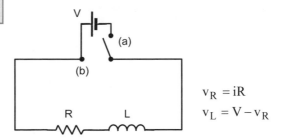

$$v_R = iR$$
$$v_L = V - v_R$$

Let connection (a) be made. Because of the self induced Emf, the current (and hence v_R) in the circuit does not rise to its final value at the instant the circuit is closed, but grows at a rate which depends on the **inductance** (Henrys) and resistance (Ohms) of the circuit.

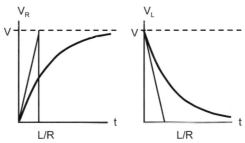

Discharging

$$0 = v_R + v_L$$

$$= iR + L\frac{di}{dt}$$

⬇

$$i = Ie^{-\frac{Rt}{L}}$$

$$v_R = Ve^{-\frac{Rt}{L}}$$

$$v_L = -Ve^{-\frac{Rt}{L}}$$

When connection (b) is made the current (and v_R) does not fall to zero immediately but falls at a rate which depends on L and R. The energy required to maintain the current during the decay is provided by the energy stored in the magnetic field of the conductor

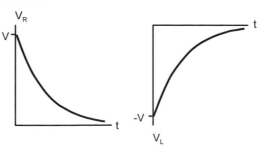

3.5.6 R-L filter circuits

Low pass (Choke) circuit

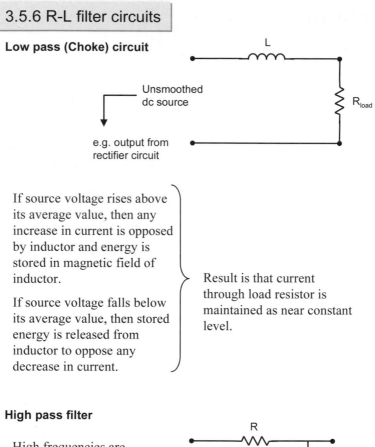

If source voltage rises above its average value, then any increase in current is opposed by inductor and energy is stored in magnetic field of inductor.

If source voltage falls below its average value, then stored energy is released from inductor to oppose any decrease in current.

Result is that current through load resistor is maintained as near constant level.

High pass filter

High frequencies are blocked by the inductor, low frequencies are passed through.

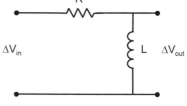

3.5.7 Integrator/Differentiator

Integrating circuit

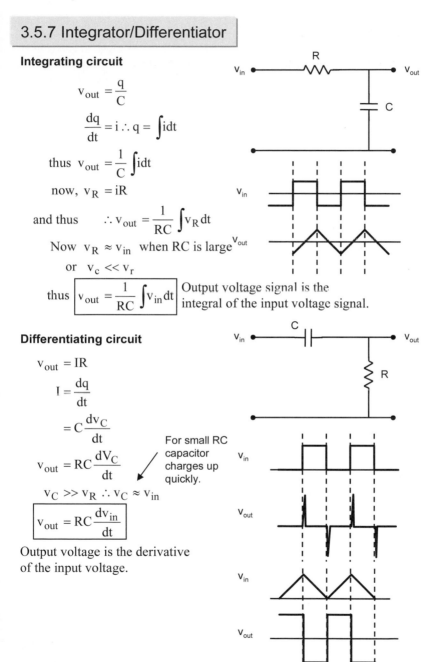

$$v_{out} = \frac{q}{C}$$

$$\frac{dq}{dt} = i \therefore q = \int i\, dt$$

thus $v_{out} = \frac{1}{C} \int i\, dt$

now, $v_R = iR$

and thus $\therefore v_{out} = \frac{1}{RC} \int v_R\, dt$

Now $v_R \approx v_{in}$ when RC is large

or $v_c \ll v_r$

thus $\boxed{v_{out} = \frac{1}{RC} \int v_{in}\, dt}$ Output voltage signal is the integral of the input voltage signal.

Differentiating circuit

$$v_{out} = IR$$

$$I = \frac{dq}{dt}$$

$$= C\frac{dv_C}{dt}$$

$$v_{out} = RC\frac{dV_C}{dt}$$

For small RC capacitor charges up quickly.

$$v_C \gg v_R \therefore v_C \approx v_{in}$$

$$\boxed{v_{out} = RC\frac{dv_{in}}{dt}}$$

Output voltage is the derivative of the input voltage.

3.5.8 Example

1. A conventional ignition system in a motor vehicle consists of an
    induction coil, the current to which is periodically switched on and off
    through mechanical "contact breaker points". A high voltage is induced
    in the secondary side of the coil when the contact breaker opens and
    closes. If the resistance of the primary side of the circuit is 8 Ω and the
    inductance of the coil is 2.4 H, calculate the following quantities:

    (a) The initial current when the contact breaker is just closed.
    (b) The initial <u>rate of change</u> of current when the contact breaker is
        just closed.
    (c) The final steady-state current
    (d) The time taken for the current to reach 95% of its maximum value

Solution:

$$i = I\left(1 - e^{-\frac{R}{L}t}\right)$$

$@t = 0, i = 0$

$V = IR$

$12 = I8$

$I = 1.5\,A$

$$\frac{di}{dt} = I\frac{R}{L}e^{-\frac{R}{L}t}$$

$@t = 0$

$$\frac{di}{dt} = \frac{R}{L}I$$

$$= \frac{8(1.5)}{2.4}$$

$$= 5\,A\,s^{-1}$$

$$0.95(1.5) = 1.5\left\{1 - e^{-\frac{8}{2.4}t}\right\}$$

$$0.05 = e^{-3.33}t$$

$$t = 0.9\,s$$

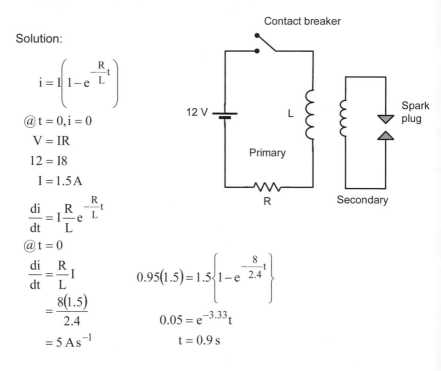

Contact breaker

12 V   L   Primary   R

Spark plug

Secondary

3.6 AC Circuits

Summary

$$v = V_o \sin(\omega t)$$ Rms voltage

$$V_{rms} = \frac{V_o}{\sqrt{2}} = 0.707 V_p$$ Rms voltage

$$I_{rms} = \frac{I_o}{\sqrt{2}} = 0.707 I_p$$ Rms current

$$X_C = \frac{1}{\omega C}$$ Capacitive reactance

$$X_L = \omega L$$ Inductive reactance

$$P_R = V_{rms} I_{rms}$$ Reactive power

$$P_{av} = V_{rms} I_{rms} \cos \phi$$ Average (active) power

$$S = V_{rms} I_{rms}$$ Apparent power

$$|Z| = \sqrt{R^2 + (X_L - X_C)^2}$$
$$\tan \phi = \left[\frac{X_L - X_C}{R} \right]$$ Impedance

$$\frac{V_{out}}{V_{in}} = \frac{1}{\sqrt{1 + R^2 \omega^2 C^2}}$$ Low pass filter

$$\frac{V_{out}}{V_{in}} = \frac{R\omega C}{\sqrt{R^2 \omega^2 C^2 + 1}}$$ High pass filter

$$R\omega C = 1$$ 3db point

3.6.1 AC Voltage

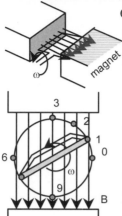

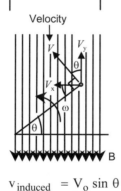

Consider a constant angular speed (ω)
- at (0) motion of conductor is parallel to B, hence induced voltage = 0
- at (1), conductor has begun to cut magnetic field lines B, hence some voltage is induced.
- at (2), conductor cuts magnetic field lines at a greater rate than (1) and thus a greater voltage is induced.
- at (3), conductor cuts magnetic field lines at maximum rate, thus maximum voltage is induced.
- From (3) to (6), the rate of cutting becomes less.
- At (6), conductor moves parallel to B and v = 0.
- From (6) to (9), conductor begins to cut field lines again but in the opposite direction, hence, induced voltage is reversed in polarity.

The **induced voltage** is directly proportional to the rate at which the conductor cuts across the magnetic field lines. Thus, the induced voltage is proportional to the velocity of the conductor in the x direction ($V_x = V \sin\theta$). The velocity component V_y is parallel to the field lines and thus does not contribute to the rate of "cutting".

Velocity

RH rule:
- fingers: direction of field
- thumb: direction of Vx
- palm: force on *positive* charge carriers

Thus, current is coming out from the page.

$$V_{induced} = V_o \sin\theta$$

⮡ maximum (peak) voltage V_o induced at $\theta = \pi/2$

V_p depends on:
- total number of flux lines through which the conductor passes
- angular velocity of loop
- no. turns of conductor in loop

since $\omega = \dfrac{\theta \text{ radians}}{t}$

then $\boxed{v = V_o \sin(\omega t)}$

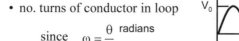

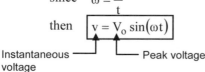

Instantaneous voltage ——   —— Peak voltage

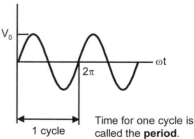

Time for one cycle is called the **period**.

3.6.2 Resistance

The **instantaneous voltage** across the resistor is:

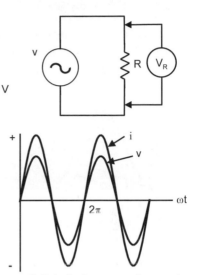

$$v_r = V_o \sin \omega t$$

└──► Maximum value of V

The **instantaneous current** in the resistor is:

$$i = \frac{V}{R}$$

$$= \frac{V_o}{R} \sin \omega t$$

The maximum current in the resistor is when $\sin \omega t = 1$ thus:

$$I_o = \frac{V_o}{R}$$

$$\therefore i = I_o \sin \omega t$$

Both instantaneous voltage and current are functions of (ωt). Thus, they are in "phase".

Instantaneous power:

$$p = vi$$

$$= i^2 R$$

$$= (I_o \sin \omega t)^2 R$$

$$= I_o^2 R \sin^2 \omega t$$

$$P_o = I_o^2 R$$

$$\boxed{p = P_o \sin^2 \omega t}$$

- power is a function of $\sin^2 \omega t$
- power is sinusoidal in nature with a frequency of twice the instantaneous current and voltage and is always positive indicating power continuously supplied to the resistor.

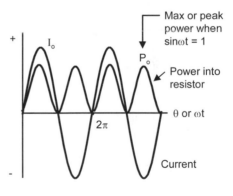

Max or peak power when $\sin \omega t = 1$

Power into resistor

Current

3.6.3 rms voltage and current

The area under the power vs time function is energy. Thus, it is possible to calculate an **average power** level which, over one cycle, is associated with the amount of energy carried in one cycle of alternating power.

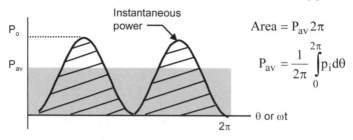

Area $= P_{av} 2\pi$

$$P_{av} = \frac{1}{2\pi} \int_0^{2\pi} p_i d\theta$$

This energy would be that given by an equivalent dc, or steady-state, voltage and current over a certain time period compared to that from an alternating current and voltage for the same time period.

Average power:

$$P_{av} = \frac{\int_0^{2\pi} p_i d\theta}{2\pi}$$

$$= \frac{1}{2\pi} \int_0^{2\pi} i^2 R d\theta$$

$$= \frac{R}{2\pi} \int_0^{2\pi} i^2 d\theta$$

$$= \frac{R}{2\pi} \int_0^{2\pi} I_o^2 \sin^2 \theta d\theta$$

$$= \frac{I_o^2 R}{2\pi} \int_0^{2\pi} \sin^2 \theta d\theta$$

this integral evaluates to just π

$$= \frac{I_o^2 R}{2}$$

Now: $P_{av} = \dfrac{I_o^2}{2} R$

What equivalent **steady-state** current would give the same average power as an alternating current?

Let: $I_{rms} = \dfrac{I_o}{\sqrt{2}}$ This result is only for sinusoidal signals.

Thus: $P_{av} = I_{rms}^2 R$ For Resistor circuit only. See later for LCR series circuit.

or

$$P_{av} = I_{rms} V_{rms}$$

↑ ↑

Equivalent steady-state values which give the same power dissipation as the application of an alternating current with peak values V_p and I_p

$$V_{rms} = \frac{V_0}{\sqrt{2}} \qquad I_{rms} = \frac{I_0}{\sqrt{2}}$$

$$= 0.707 V_0 \qquad = 0.707 I_0$$

In AC circuits, V and I without subscripts indicate rms values unless stated otherwise.

3.6.4 Capacitive reactance

The AC source supplies an alternating voltage v. This
voltage appears across the capacitor.

In general, $C = \dfrac{Q}{V}$ (q,v are instantaneous
values and thus
functions of t)

thus $q = Cv$

$\dfrac{dq}{dt} = C\dfrac{dv}{dt}$ differentiating
w.r.t. time

$i = C\dfrac{dv}{dt}$ C is a "constant"

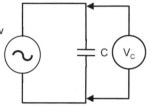

Instantaneous current is proportional to the rate of change of voltage.

The instantaneous current is a
maximum I_p when the rate of
change of voltage is a
maximum. Also, the
maximum voltage V_p
only appears across the
capacitor *after* it has
become charged
whereupon the current
I drops to zero. Thus,
maximums and minimums in
the instantaneous current lead
the maximums and minimums
in the instantaneous voltage by
$\pi/2$.

Maximums in
current in
capacitor
precede
maximums in
voltage across it.

Now, $v = V_o \sin(\omega t)$

$i = C\dfrac{d}{dt}V_o \sin(\omega t)$ since $i = C\dfrac{dv}{dt}$

thus $i = \omega C V_o \cos(\omega t)$

$= [\omega C V_o]\sin\left(\omega t + \dfrac{\pi}{2}\right)$

$I_o = \omega C V_o$ @ $i = I_p$

$i = I_o \sin\left(\omega t + \dfrac{\pi}{2}\right)$

$\dfrac{1}{\omega C} = \dfrac{V_o}{I_o}$ Can be peak or rms
but not instantaneous

$= X_C$

Capacitive
reactance (Ω)

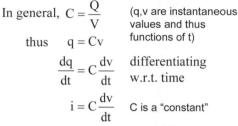

Capacitive reactance is the opposition to
alternating current by capacitance. The
opposition tendered depends upon the rate of
change of <u>voltage</u> through the circuit.

What is capacitive reactance? How can a capacitor offer a **resistance** to alternating current.

Consider a capacitor connected to a DC supply so that the polarity of the applied voltage can be reversed by a switch.

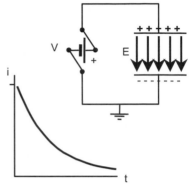

When the switch is first closed, it takes time for the charge Q to accumulate on each plate. Charge accumulation proceeds until the voltage across the capacitor is equal to the voltage of the source. During this time, current flows in the circuit.

When the polarity is reversed, the capacitor initially discharges and then charges to the opposite polarity. Current flows in the opposite direction while reverse charging takes place until the voltage across the capacitor becomes equal to the supply voltage whereupon current flow ceases.

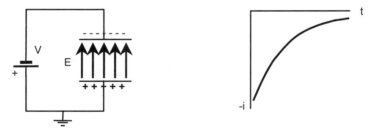

Now, if the switch were to be operated very quickly, then, upon charging, the current would not have time to drop to zero before the polarity of the supply voltage was reversed. Similarly, on reverse charging, the reverse current would not have time to reach zero before the polarity of the source was reversed. Thus, the current would only proceed a short distance along the curves as shown and a continuous alternating current would result. The faster the switch over of polarity, the greater the average or rms ac current. Thus, the "resistance" to ac current is greater at lower frequencies and lower at high frequencies.

3.6.5 Inductive reactance

Let the inductor have no resistance. Thus, any voltage that appears across the terminals of the inductor must be due to the self-induced voltage in the coil by a changing current through it (self-inductance).

At any instant, $v_L + v = 0$ by Kirchhoff

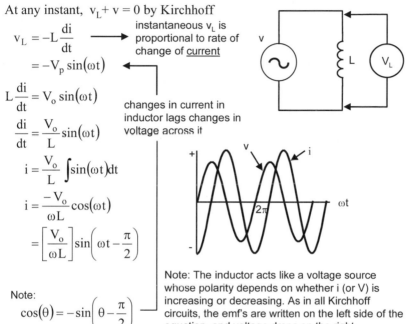

$$v_L = -L\frac{di}{dt}$$

instantaneous v_L is proportional to rate of change of current

$$= -V_p \sin(\omega t)$$

$$L\frac{di}{dt} = V_o \sin(\omega t)$$

changes in current in inductor lags changes in voltage across it

$$\frac{di}{dt} = \frac{V_o}{L}\sin(\omega t)$$

$$i = \frac{V_o}{L}\int \sin(\omega t)dt$$

$$i = \frac{-V_o}{\omega L}\cos(\omega t)$$

$$= \left[\frac{V_o}{\omega L}\right]\sin\left(\omega t - \frac{\pi}{2}\right)$$

Note:
$$\cos(\theta) = -\sin\left(\theta - \frac{\pi}{2}\right)$$

Note: The inductor acts like a voltage source whose polarity depends on whether i (or V) is increasing or decreasing. As in all Kirchhoff circuits, the emf's are written on the left side of the equation, and voltage drops on the right.

Now, i will be a maximum I_p when $\sin\left(\omega t - \frac{\pi}{2}\right) = 1$

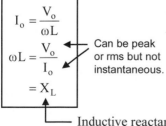

$$I_o = \frac{V_o}{\omega L}$$

$$\omega L = \frac{V_o}{I_o}$$

Can be peak or rms but not instantaneous.

$$= X_L$$

Inductive reactance (Ω)

Inductive reactance is the opposition to alternating current by inductance. The opposition tendered depends upon the rate of change of current through the circuit.

For high frequencies, the magnitude of the induced back emf is large and this restricts the maximum current that can flow before the polarity of the voltage changes over. Thus, the reactance increases with increasing frequency.

3.6.6 Reactive power (capacitor)

For a capacitor:
$$v = V_o \sin(\omega t)$$
$$i = I_o \cos(\omega t)$$

In the case of a resistor, the power dissipated $P_{av} = V_{rms}I_{rms}$ and is the average of the **instantaneous power** fluctuations over one complete cycle. However, for a capacitor, the average of the instantaneous power fluctuations is now zero.

Instantaneous power:

$$p = vi$$
$$= V_o I_o \sin \omega t \cos \omega t$$
$$= \frac{1}{2} V_o I_o \sin 2\omega t$$

Average power:
$$P_{av} = \frac{1}{2\pi} \int_0^{2\pi} vi \, d\theta$$

$$= \frac{V_o I_o}{2\pi} \int_0^{2\pi} \sin \theta \cos \theta \, d\theta$$

$$= \frac{V_o I_o}{4\pi} \left[\sin^2 \theta \right]_0^{2\pi}$$

$$= 0$$

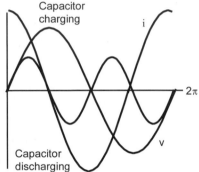

For a capacitor, the product $V_{rms}I_{rms}$ is called the **reactive power**. The reactive power is *an indication* of the power alternately supplied and discharged from the capacitor.

$$\boxed{P_R = V_{rms}I_{rms}}$$

The reactive power is *directly proportional* to the power supplied to and obtained from the capacitor.

Volts Amps Reactive

Reactive power is given the units **var** to distinguish this type of power from **Watts** which is reserved for dissipative or **active power**.

3.6.7 Reactive power (inductor)

For an inductor: $v = V_o \sin(\omega t)$

$i = -I_o \cos(\omega t)$

Instantaneous power

$$p = vi$$

$$= -V_o I_o \sin \omega t \cos \omega t$$

$$= -\frac{1}{2} V_o I_o \sin 2\omega t$$

In the case of a resistor, the power dissipated $P_{av} = V_{rms} I_{rms}$ and is the average of the instantaneous power fluctuations. However, for an inductor, the average of the instantaneous power fluctuations is now zero.

Average power
$$P_{av} = \frac{1}{2\pi} \int_0^{2\pi} vi \, d\theta$$

$$= \frac{-V_o I_o}{2\pi} \int_0^{2\pi} \sin \theta \cos \theta \, d\theta$$

$$= \frac{-V_o I_o}{4\pi} \left[\sin^2 \theta\right]_0^{2\pi}$$

$$= 0$$

$\Theta = \omega t$

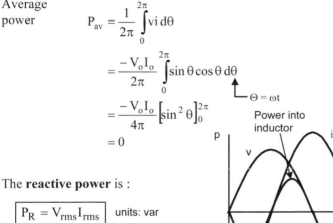

Power into inductor

Power out of inductor

The **reactive power** is :

$$\boxed{P_R = V_{rms} I_{rms}} \quad \text{units: var}$$

and *is an indication of* the power alternately supplied to, and obtained from, an inductive reactance. If there is some resistance in the circuit, such as from the windings of the inductor, then energy loss occurs and there is some **active power** involved.

3.6.8 LCR series circuit

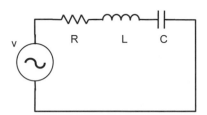

A varying voltage v from the source will cause a varying instantaneous current i to flow in the circuit. Because it is a series circuit, the current must be the same in each part of the circuit at any particular time t.

- For the resistance, changes in v_R are in phase with those of i
- For the inductor, changes in v_L precedes those of i by $\pi/2$
- For the capacitor, changes in v_C follow those of i by $\pi/2$

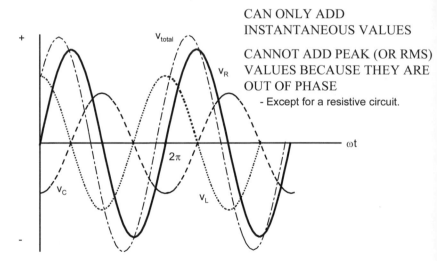

CAN ONLY ADD INSTANTANEOUS VALUES

CANNOT ADD PEAK (OR RMS) VALUES BECAUSE THEY ARE OUT OF PHASE
- Except for a resistive circuit.

From Kirchoff, $v_{Total} = v_R + v_L + v_C$ at any instant. Note that each of these voltages do not reach their peak values when V_{Total} reaches a maximum, thus $|V_{oTotal}| \lessgtr |V_{oR}| + |V_{oL}| + |V_{oC}|$. Also, since the rms value of any voltage = 0.707 V_o, then $|Vrms_{total}| \lessgtr |Vrms_R| + |Vrms_L| + |Vrms_C|$

Algebraic addition generally only applies to instantaneous quantities (can be applied to other quantities, e.g. peak or rms, if in current and voltage are in phase - such as resistor circuit only).

Note also that the resultant of the addition of sine waves of same frequency results in a sine wave of same frequency.

3.6.9 LCR circuit – peak and rms voltage

In LCR series circuits, how then may the total or <u>resultant</u> peak (or rms) values of voltage and current be determined from the individual peak (or rms) voltages? A VECTOR approach is needed (can use **complex numbers**).

Consider the axes below which indicates either peak (or rms) voltages V_R, V_C and V_L:

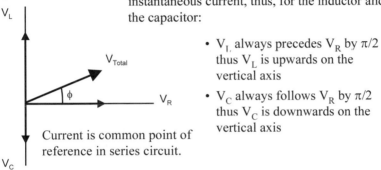

In a *series* circuit, the instantaneous voltage across the resistor is always in phase with the instantaneous current, thus, for the inductor and the capacitor:

- V_L always precedes V_R by $\pi/2$ thus V_L is upwards on the vertical axis
- V_C always follows V_R by $\pi/2$ thus V_C is downwards on the vertical axis

Current is common point of reference in series circuit.

The resultant peak or rms voltage V_T is the <u>vector</u> sum of $V_R + V_L + V_C$

The angle ϕ is the phase angle of the resultant peak (or rms) voltage w.r.t. the peak (or rms) common current and is found from:

$$\tan \phi = \frac{V_L - V_C}{V_R}$$

For an AC series circuit:
- same current flows in all components.
- vector sum of rms or peak voltages must equal the applied rms or peak voltage
- algebraic sum of instantaneous voltages equals the applied instantaneous voltage

In **complex number** form:

$$V_T = V_R + j(V_L - V_C)$$

Complex numbers are a convenient mathematical way to keep track of directions or "phases" of quantities.

3.6.10 Power in LCR circuit

For a resistor: $p = P_o \sin^2(\omega t)$

$$= V_o I_o \sin^2(\omega t)$$

However, <u>in general</u>, there is a phase
difference between i and v :

Note: no phase
difference for resistor on
its own.

$$i = I_o \sin(\omega t)$$
$$v = V_o \sin(\omega t + \phi)$$

where ϕ may be either positive (capacitor) or negative (inductor).

The instantaneous power in circuit containing resistance and
reactance:

$$p = vi$$
$$= V_o I_o \sin \omega t \sin(\omega t + \phi)$$

Instantaneous
power for: ⌐

$$= V_o I_o \sin \omega t (\sin \omega t \cos \phi + \cos \omega t \sin \phi)$$

$$p = \cos \phi \underbrace{(V_o I_o \sin^2 \omega t)}_{} + \sin \phi \underbrace{(V_o I_o \sin \omega t \cos \omega t)}_{}$$

Instantaneous ——
power for resistor

Capacitor; p_C

or

$$(-V_o I_o \sin \omega t \cos \omega t)$$

Inductor, p_L

Average power for the circuit:

$$P_{av} = \frac{1}{2\pi} \int_0^{2\pi} p\, d\omega t$$

The average power for the
capacitor and the inductor = 0,
thus, this second integral = 0.

$$= \frac{V_o I_o \cos \phi}{2\pi} \int_0^{2\pi} \sin^2 \omega t \, d\omega t + \sin \phi \int_0^{2\pi} [p_C ; p_L] d\omega t$$

$$= \frac{V_o I_o \cos \phi}{2}$$

$$= \frac{V_o}{\sqrt{2}} \frac{I_o}{\sqrt{2}} \cos \phi$$

$$\boxed{P_{av} = V_{rms} I_{rms} \cos \phi}$$

Cos ϕ is called the
power factor.

units: Watts

This represents the power dissipated in a resistor
in an LCR circuit or consumed in some other
fashion (eg to drive an electric motor).

Because P_{av} is the
power that is actually
used by the circuit, it is
called the **real** or **active**
power.

p | into circuit
P_{av}
-p | back from circuit

3.6.11 Apparent power

Consider an LCR circuit in which the rms voltage across the source and the (common) current are measured with meters.

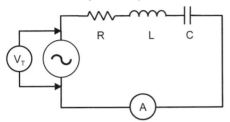

Even if we know V_{rms} and I_{rms}, we cannot determine the average (ie. the active) power consumed by this circuit since we need to also measure $\cos \phi$

$$P_{av} = V_{rms} I_{rms} \cos \phi$$

But, knowing the rms voltage across the resistor, inductor and capacitor, then the angle ϕ *can* be obtained from:

$$\tan \phi = \frac{V_L - V_C}{V_R} \longrightarrow \text{Peak or rms values}$$

By Kirchhoff's law, we know that the peak or rms value of the applied voltage is the vector sum of the peak or rms voltage across each of the components. Thus, the magnitude of the peak or rms applied voltage V_T is (by Pythagoras):

$$|V_T| = \sqrt{|V_R|^2 + |V_L - V_C|^2}$$

A similar analysis applies to parallel circuits where voltage is the common point of reference.

Multiply both sides by $|I|$

$$|V_T||I| = \sqrt{|V_R|^2|I|^2 + |V_L - V_C|^2|I|^2}$$

$\qquad\qquad\qquad$ active $\qquad\qquad$ reactive

Thus, if $|V|$ and $|I|$ refer to rms values, then the power as calculated by the product of measured values of $V_{rms} I_{rms}$ is called the **apparent power** S and is a combination of the active and reactive power within the circuit.

$$S = V_{rms} I_{rms} \qquad \text{units: VA}$$

This formula applies to series and parallel circuits.

The **apparent power** is the total average power that needs to be supplied by the power source (ie the energy company). Some of this power ($\cos\phi$) is used by the circuit (the **active power**). Some of this power ($\sin\phi$ - the **reactive power**) is alternately supplied to and received from the circuit as the reactive components alternately charge and discharge.

Note: The apparent power S is not simply the sum of P_{av} and P_R due to the phase difference between V and I.

Although the reactive power is not consumed by the circuit, associated transmission losses in alternate transfers of energy lead to undesirable inefficiencies.

3.6.12 Power factor

The **active power** is the average power actually consumed or dissipated by an LCR circuit.

$$P_{av} = V_{rms}I_{rms}\cos\phi$$

The **apparent power** is that which must be supplied to the circuit.

$$S = V_{rms}I_{rms}$$

Thus, the **power factor** is: $\cos\phi = \dfrac{P_{av}}{S}$

The **reactive power** is the power alternately exchanged to and from reactive components:

$$P_{reactive} = V_{rms}I_{rms}\sin\phi$$

I_{rms} is the total current

Even though an LCR circuit only uses a portion of the power supplied to it, it is desirable to reduce the transmission losses associated with energy transfers to and from the reactive parts of the circuit (i.e. a power factor of 100% is desirable).

In an inductively reactive circuit the phase angle ϕ is negative and the **power factor** is termed **lagging**. In a capacitively reactive circuit, the phase angle is positive and the power factor is termed **leading**. Most industrial circuits are either capacitively or inductively reactive. To reduce transmission losses, capacitors and inductors may be purposely added to the overall circuit to ensure that the power factor is as close to 100% as possible.

Power factor correction involves connecting either a capacitor or inductor in parallel with the load so as to bring the power factor to 100%. When this occurs, the additional energy alternately transferred to and from the source is now exchanged between the load and the power factor correction component.

Example: Consider a 1 kW (active) load with a 70% lagging (i.e. inductively reactive) power factor. From the point of view of the source, the apparent power is:

$$S = \frac{1000}{0.7} \qquad 0.7 = \cos^{-1}\phi$$
$$= 1428 \text{ VA} \quad \phi = -45.6°$$

For a *lagging* power factor, we require a *capacitor* to be connected in parallel with it to bring the power factor up to 100%. What value of capacitor is required? Since, in this example, the circuit is inductively reactive, it is simply necessary to use a capacitor whose reactive power is equal to the reactive power of the inductively reactive load

$$\begin{aligned} P_R &= V_{rms}I_{rms} \\ &= \frac{V_{rms}^2}{X_C} \\ &= V_{rms}^2\omega C \end{aligned}$$

$$P_R = V_{rms}I_{rms}\sin\phi$$

where V_{rms} is the rms voltage <u>across the capacitor</u> (same as voltage across load since capacitor is in parallel with it) and I_{rms} is the rms current (as yet unknown) through the parallel capacitor.

3.6.13 Impedance

The total opposition to current in an AC circuit is called **impedance**. For a series circuit, the impedance is the vector sum of the resistances and the reactances within the circuit

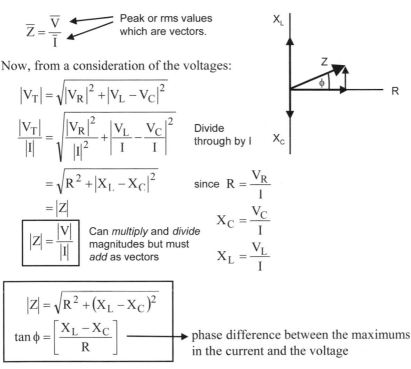

$$\overline{Z} = \frac{\overline{V}}{\overline{I}}$$

Peak or rms values which are vectors.

Now, from a consideration of the voltages:

$$|V_T| = \sqrt{|V_R|^2 + |V_L - V_C|^2}$$

$$\frac{|V_T|}{|I|} = \sqrt{\frac{|V_R|^2}{|I|^2} + \left|\frac{V_L}{I} - \frac{V_C}{I}\right|^2}$$ Divide through by I

$$= \sqrt{R^2 + |X_L - X_C|^2}$$ since $R = \dfrac{V_R}{I}$

$$= |Z|$$

$$|Z| = \frac{|V|}{|I|}$$ Can *multiply* and *divide* magnitudes but must *add* as vectors

$$X_C = \frac{V_C}{I}$$

$$X_L = \frac{V_L}{I}$$

$$|Z| = \sqrt{R^2 + (X_L - X_C)^2}$$

$$\tan \phi = \left[\frac{X_L - X_C}{R}\right]$$ ⟶ phase difference between the maximums in the current and the voltage

For an RC series circuit:

$$\tan \phi = \left[\frac{-X_C}{R}\right]$$

$$= \frac{1}{R\omega C}$$

For an RL series circuit:

$$\tan \phi = \left[\frac{X_L}{R}\right]$$

$$= \frac{\omega L}{R}$$

If $X_C > X_L$, circuit is **capacitively reactive**.
If $X_C < X_L$, circuit is **inductively reactive**.
If $X_C = X_L$ - then **resonance**.

3.6.14 Example

1. Determine the expression for V_{out}/V_{in} for the following high pass filter circuit. Also determine the condition for $\dfrac{V_{out}}{V_{in}} = \dfrac{1}{\sqrt{2}}$

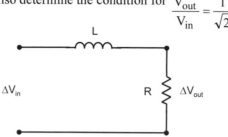

Solution:

$$V_{in} = IZ$$

$$= I\sqrt{R^2 + X_L^2}$$

$$V_{out} = IR$$

$$\frac{V_{out}}{V_{in}} = \frac{IR}{I\sqrt{R^2 + \omega^2 L^2}}$$

$$= \frac{1}{\sqrt{1 + \dfrac{\omega^2 L^2}{R^2}}}$$

$$\frac{1}{\sqrt{2}} = \frac{1}{\sqrt{1 + \dfrac{\omega^2 L^2}{R^2}}}$$

$$\frac{1}{2} = \frac{1}{1 + \dfrac{\omega^2 L^2}{R^2}}$$

$$\frac{\omega^2 L^2}{R^2} = 1$$

$$\frac{\omega L}{R} = 1$$ This condition is known as the 3 db point and is often used for defining the frequency range or bandwidth of the filter.

3.7 Electromagnetic waves

Summary

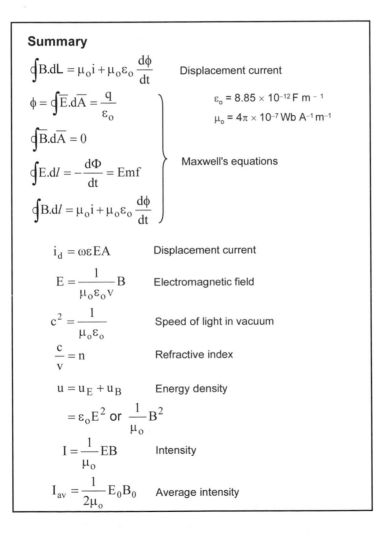

$$\oint B.dL = \mu_0 i + \mu_0 \varepsilon_0 \frac{d\phi}{dt}$$ Displacement current

$$\phi = \oint \overline{E}.d\overline{A} = \frac{q}{\varepsilon_0}$$ $\varepsilon_0 = 8.85 \times 10^{-12}\, F\, m^{-1}$

$\mu_0 = 4\pi \times 10^{-7}\, Wb\, A^{-1}\, m^{-1}$

$$\oint \overline{B}.d\overline{A} = 0$$

$$\oint E.dl = -\frac{d\Phi}{dt} = Emf$$ Maxwell's equations

$$\oint B.dl = \mu_0 i + \mu_0 \varepsilon_0 \frac{d\phi}{dt}$$

$$i_d = \omega \varepsilon EA$$ Displacement current

$$E = \frac{1}{\mu_0 \varepsilon_0 v} B$$ Electromagnetic field

$$c^2 = \frac{1}{\mu_0 \varepsilon_0}$$ Speed of light in vacuum

$$\frac{c}{v} = n$$ Refractive index

$$u = u_E + u_B$$ Energy density

$$= \varepsilon_0 E^2 \text{ or } \frac{1}{\mu_0} B^2$$

$$I = \frac{1}{\mu_0} EB$$ Intensity

$$I_{av} = \frac{1}{2\mu_0} E_0 B_0$$ Average intensity

3.7.1 Charging a capacitor

Consider an **air-filled capacitor** connected to a DC supply through a switch. When the switch is first closed, it takes time for the charge Q to accumulate on each plate. Charge accumulation proceeds until the voltage across the capacitor is equal to the voltage of the source. During this time, current flows in the wire. Note, no current passes through the air space between the plates (since there are no mobile charge carriers there).

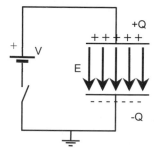

Now, let us apply **Ampere's law** to various parts of the circuit. While there is a current flowing in the wire, there is of course a magnetic field which encircles the wire the strength of which can be calculated according to Ampere's law.

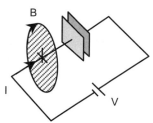

$$\oint B.dl = \mu_0 I$$

The line integral of B around the closed path is proportional to the current I which "pierces" the surface bounded by the path. Note that I flows through the flat surface (shaded) enclosed by the circle B.

Now, what happens if we stretch the surface so that it lies within the gap of the capacitor? The current which pierces the surface is now zero, which implies that the magnetic field encircling the wire is now zero!! Something must be wrong with Ampere's law!

It shouldn't matter which surface we choose! The line integral (around the line which defines the surface) comes out the same. Therefore there must be a current I!! But how can there be a current through the air in the gap ???

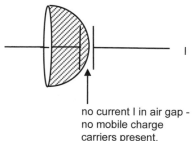

no current I in air gap - no mobile charge carriers present.

3.7.2 Displacement current

Maxwell argued that although there is no actual **conventional current** in the gap (since the material inside the gap is an insulator - eg. air, with no mobile charge carriers to "carry" the current), one could imagine that there is indeed a current arising from the small displacement of charges within the atoms and molecules of the gap material.

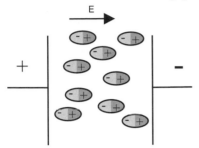

The amount of lining up of molecules in the insulating material (or "dielectric") is characterised by the permittivity of the material. Materials with a high permittivity have molecules that line up rather easily in the electric field. When this alignment occurs, a series of internal fields are created which has the effect of reducing the total field between the plates.

Here's how it works:

1. As current flows in the wires, build up of charge on either plate of the capacitor establishes a field E within the gap. This field starts from zero and reaches a maximum when the capacitor is fully charged.

2. The field E acts upon the air molecules within the gap causing them to become **polarised** as electrons within the atoms are attracted towards the positive plate, and positively charged nuclei become attracted to the negative plate.

3. The charges within the molecules are **displaced** but cannot move very far. The charges actually move while the field E is increasing (or decreasing). When the field E is steady, the charges remain stationary in their displaced positions.

4. Any movement of charge is an electric current, thus, when the field E is changing, there is a current in the gap as the charges within the molecules are displaced from further from their equilibrium positions. This current is called the **displacement current**. The displacement current only flows when the field E is either increasing or decreasing.

3.7.3 Ampere's law

Maxwell argued further that the effects of a **displacement current** exist even when there is a vacuum between the plates! He reasoned that the changing electric field E between the plates results in there being an encircling magnetic field within the gap - one didn't actually require there to be a movement of charges to produce a magnetic field, all that was needed was a changing electric field!

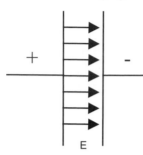

The changing electric field within the gap is electrically and magnetically equivalent to the existence of a displacement current within the gap.

As capacitor charges up, electric field E in the gap increases. The electric field is provided by the build up of charge on the plates of the capacitor rather than the presence of actual charge carriers in motion.

Thus, for the example of a charging capacitor where there is a changing current in a circuit, Ampere's law still holds as long as the displacement current i_d is included:

$$\oint B.dl = \mu_o(i + i_d)$$

Note the use of lower case i to signify changing or "dynamic" quantities.

If we choose a surface which is pierced by the wire (i.e. conductor) then the current is just i and $i_d = 0$. If we choose a surface which is pierced by the field E, then the current is i_d and i = 0. In any case, $i = i_d$.

Now, consider the electric flux ϕ between the plates.

$$\phi = \oint \overline{E}.d\overline{A}$$

$$= EA \quad \text{for a constant E over area A}$$

$$q = \varepsilon_o \phi \quad \text{by Gauss' law}$$

$$= \varepsilon_o EA$$

$$i_d = i = \frac{dq}{dt}$$

$$= \varepsilon_o \frac{d\phi}{dt}$$

This equation states that a magnetic field B may be created by either a current i (steady or changing) and/or a changing electric flux.

Thus, Ampere's law becomes:

$$\oint B.dl = \mu_o\left(i + \varepsilon_o \frac{d\phi}{dt}\right)$$

Conduction current   Displacement current

For parallel plates of area A and gap d, the displacement current is: $i_d = \omega \varepsilon EA$

3.7.4 Maxwell's equations

Let us summarise the various equations of electricity and magnetism:

Gauss' law:
(Electric charge) $\phi = \oint \overline{E.dA}$

$$= \frac{q}{\varepsilon_o}$$

These four equations are known as **Maxwell's equations** and can be used to quantify all aspects of electricity and magnetism, including the existence and properties of electromagnetic waves.

Gauss' law:
(Magnetism)

$$\oint \overline{B.dA} = 0$$

Faraday's law:

$$\oint E.dl = -\frac{d\Phi}{dt}$$

$$= Emf$$

Ampere's law:

$$\oint B.dl = \mu_o i + \mu_o \varepsilon_o \frac{d\phi}{dt}$$

Note the similarity between Ampere's law and Faraday's law. These laws are not quite symmetric. Ampere's law says that a magnetic field can be created by either a steady (or changing) current in a conductor, or a changing electric field in space. Faraday's law says that an electric field can only be created by a changing magnetic field. This asymmetry arises because of the fact that magnetic field lines always form closed loops. An electric field line may originate on one isolated charge and terminate on another. Magnetic field lines always close upon themselves. There is no such thing as an isolated magnetic "charge". The fact that isolated electric charges can exist means that an electric current is a physical possibility as the charges move from one place to another. This leads to the $\mu_o i$ term in Ampere's law. The impossibility of any "magnetic" current means that this term is missing in Faraday's law. In fact, this is exactly what the first two equations are saying. In the electric field, field lines may originate from a charged particle within the surface and then pass out through a surface and terminate on isolated charges some distance away. Gauss' law for magnetism says that for any closed surface in space, the total magnetic flux is zero meaning that all magnetic flux lines join up with themselves.

3.7.5 Electric and magnetic fields

Consider an electric field which leaves the charge which created it and moves through space with a velocity v. The field is on the point of entering an area A.

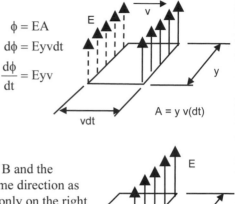

In a time interval dt, the field has travelled a distance vdt. The electric flux through the area A has changed.

$$\phi = EA$$
$$d\phi = Eyvdt$$
$$\frac{d\phi}{dt} = Eyv$$

$A = y\,v(dt)$

Now, consider Ampere's law:

$$\oint B.dl = \mu_0\varepsilon_0 \frac{d\phi}{dt}$$
$$= \mu_0\varepsilon_0 Eyv$$

The line integral is the product of B and the length y (since y lies along the same direction as B and after time dt, B now exists only on the right hand edge of the area), thus:

$$\oint B.dl = \mu_0\varepsilon_0 \frac{d\phi}{dt}$$
$$By = \mu_0\varepsilon_0 Eyv$$
$$B = \mu_0\varepsilon_0 Ev$$
$$\boxed{E = \frac{1}{\mu_0\varepsilon_0 v}B}$$

Note: we are considering the movement of the field only, there are no charges present hence no I term in Ampere's law.

Consider now the passage of a magnetic field B through an area A.

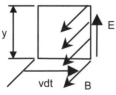

The change in magnetic flux is:

$$\Phi = BA$$
$$d\Phi = Byvdt$$
$$\frac{d\Phi}{dt} = Byv$$

An application of Faraday's law yields:

$$\oint E.dl = -\frac{d\Phi}{dt}$$
$$Ey = Byv$$
$$\boxed{E = vB}$$

Faraday's law and **Ampere's law** say that a moving electric field is accompanied by a magnetic field and a moving magnetic field is accompanied by an electric field. This can only happen when v = c:

$$E = \frac{1}{\mu_0\varepsilon_0 v}B = vB$$

$$c^2 = \frac{1}{\mu_0\varepsilon_0}$$

$$\boxed{c = 3\times10^8\,ms^{-1}}$$

The fields move with the **velocity of light:** $c = 3 \times 10^8\ ms^{-1}$.

$\varepsilon_0 = 8.85 \times 10^{-12}\,F\,m^{-1}$
$\mu_0 = 4\pi \times 10^{-7}\,Wb\,A^{-1}\,m^{-1}$

3.7.6 Electromagnetic waves

Previously we considered an electric field moving through space. No charges were present, hence, in **Ampere's law**, we only needed to consider the displacement current. The conduction current i was set to zero.

How can an electric field move through space?

Consider an electric charge initially at rest which then accelerates and reaches a constant velocity. An electric field (and associated increasing magnetic field) moves along in space with the charge. The fields are attached to the charge.

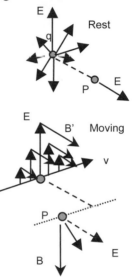

At some instant, point P, some distance from the charge, initially experiences the electric field E. When the charge q **accelerates**, the field E moves along with (i.e. it is attached to) the charge.

Some time later, the charge reaches a constant velocity and the charge and P are no longer in line with each other. The direction of E as experienced by the point P will have changed. A magnetic field also exists at point P in accordance with Ampere's law. Let us concentrate on the change in the direction of E:

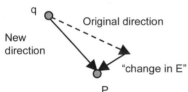

The point P does not experience the change in direction of E instantaneously. When the charge accelerates to a new velocity, it "transmits" a "change in E" vector outwards. When the point P receives this "change in E" signal, it is added to the original E at P to give the new direction of E at P.

The "change in E" signal is a E vector field which is not attached to the charge. It is a field which is "launched" into space when the charge is accelerated. The field travels through space with an accompanying magnetic field of its own at velocity c. The "change in E" signal is a moving **electromagnetic field**. A new "change in E" field is transmitted each time the charge accelerates or decelerates. **Electromagnetic waves** are travelling E and B fields in space and are created by **accelerating charges**.

3.7.7 Periodic electromagnetic waves

The most common type of travelling electromagnetic fields are those which vary continuously in a smooth periodic manner. Previously we examined an **electric field** "pulse" and its accompanying **magnetic field** pulse. Now consider a smoothly varying electric field travelling through space:

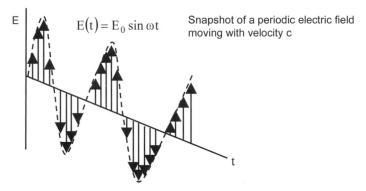

$E(t) = E_0 \sin \omega t$

Snapshot of a periodic electric field moving with velocity c

This field periodically changes in magnitude and direction.

An **electric field** travelling through space is accompanied by a magnetic field with a direction normal to the field and the velocity. Since in this example the electric field varies in magnitude periodically with time, then the magnetic field also varies periodically with time:

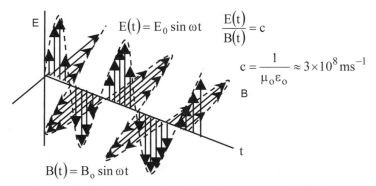

$$E(t) = E_0 \sin \omega t \qquad \frac{E(t)}{B(t)} = c$$

$$c = \frac{1}{\mu_0 \varepsilon_0} \approx 3 \times 10^8 \, \text{ms}^{-1}$$

$$B(t) = B_0 \sin \omega t$$

Because of the periodicity, the moving electric and magnetic fields are together called an **electromagnetic wave**. These "field" waves have all the properties of mechanical waves, displaying diffraction, interference, reflection and refraction. The range of frequencies most commonly encountered is termed the **electromagnetic spectrum**.

3.7.8 Electromagnetic waves in a dielectric

For an electromagnetic wave travelling in an **insulator**, or **dielectric**, the permittivity and permeability of the material has to be taken into consideration. The velocity of the wave in a dielectric becomes:

$$v = \frac{1}{\sqrt{\mu_r \varepsilon_r \mu_o \varepsilon_o}}$$

For very high frequencies (such as that of visible light) values of ε can be substantially less than the steady state value.

$$= \frac{1}{\sqrt{\varepsilon \mu}} \quad \text{where} \quad \begin{array}{l} \varepsilon = \varepsilon_r \varepsilon_o \\ \mu = \mu_r \mu_o \end{array} \quad \text{and thus:} \quad \frac{c}{v} = \sqrt{\mu_r \varepsilon_r}$$

$\mu_r \approx 1$ for most materials

$= n$ the **refractive index** of the medium.

When an electromagnetic wave is incident on a surface, part of the energy in the waves is absorbed, part of it may be reflected, and part transmitted through it.

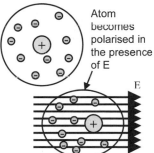

Atom becomes polarised in the presence of E

In a **dielectric**, the amount of radiation absorbed depends upon the permittivity which in general, is frequency dependent. The field E causes atoms in the material to become polarised. Losses arise as the polarisation direction changes in the presence of an oscillating field E from the wave.

For electromagnetic waves incident on a real conducting surface, *most* of the wave is reflected. That which is absorbed, diminishes in magnitude exponentially with depth into the material. The depth at which the magnitude of the fields has dropped by $\approx 63\%$ is called the **skin depth** δ.

At relatively low frequencies:

$$\delta = \frac{1}{\sqrt{\pi f \mu \sigma}}$$

At relatively high frequencies:

$$\delta = \frac{2}{\sigma} \sqrt{\frac{\varepsilon}{\mu}}$$

In a perfect conductor, the electric field is completely cancelled by the redistribution of charges which occurs and the wave is reflected.

Frequency Hz   Conductivity

Depends on frequency. The higher the frequency, the lower the value of ε.

X rays (and also gamma rays) have very short wavelengths and very high frequencies. Electrons within a material cannot respond fast enough to these frequencies to allow the usual energy dissipative mechanisms to occur. Instead, X rays lose energy by ejecting electrons from the material through which they travel (ionisation). But, the electrons most affected by this ionisation process are those in the inner most shells of an atom (K and L). Since there are relatively few electrons in these innermost shells, X rays do not have a high probability of interacting with them and hence their high penetrating ability.

3.7.9 Energy in electromagnetic waves

Consider the energy required to charge a capacitor. The energy is stored within the field between the plates of the capacitor.

$$U_E = \frac{1}{2}CV^2$$

Now, $C = \varepsilon_0 \dfrac{A}{d}$

If the space between the plates of the capacitor is filled with a dielectric, then we use $\varepsilon = \varepsilon_r \varepsilon_0$

and $V = Ed$

Thus: $U_E = \dfrac{1}{2}\varepsilon_0 \dfrac{A}{d}E^2 d^2$

$\qquad = \dfrac{1}{2}\varepsilon_0 E^2 (Ad)$

The **energy density** is the energy contained within the electric field per unit volume (J m^{-3}). Here, the volume occupied by the field E is the product Ad. Thus:

$$u_E = \frac{U_E}{Ad} = \frac{1}{2}\varepsilon E^2$$

This formula says that the energy density of an electric field is proportional to the square of the amplitude of the field.

The energy required to "charge" an inductor of length l, cross sectional area A and N number of turns is:

$$U_B = \frac{1}{2}LI^2 \qquad \text{where} \quad L = \mu_0 A \frac{N^2}{l}$$

The energy is stored with the magnetic field:

$$B = \mu_0 I \frac{N}{l}$$

$$U_B = \frac{1}{2}\left[\mu_0 A \frac{N^2}{l}\right]\left[\frac{B}{\mu_0}\frac{l}{N}\right]^2$$

$$= \frac{1}{2\mu_0}B^2 Al \longrightarrow \text{A}l \text{ is volume of field}$$

$$u_B = \frac{1}{2\mu_0}B^2$$

Energy density of a magnetic field.

Now, $E = Bc$

$$u_E = \frac{1}{2}\varepsilon_0 B^2 c^2$$

$$= \frac{1}{2}\varepsilon_0 u_B 2\mu_0 c^2$$

$$= \varepsilon_0 \mu_0 c^2 u_B$$

but $c = \dfrac{1}{\sqrt{\varepsilon_0 \mu_0}}$

$\therefore u_E = u_B$

The **total energy density** of an electromagnetic wave is thus:

$$u = u_E + u_B$$

$$= \varepsilon_0 E^2 \text{ or } \frac{1}{\mu_0}B^2$$

The energy carried by an electromagnetic wave is divided equally between the E and B fields.

3.7.10 Intensity of electromagnetic wave

The total **energy density** of an electromagnetic wave is : $u = \varepsilon_0 E^2$

The energy density is proportional to the square of the magnitude of the E field and is the energy (J) per unit volume of space containing the field.

When an electromagnetic wave travels through space, energy is transported along with it (i.e. within the field). The **intensity** is power (i.e. rate of energy transfer) transmitted per unit area.

If, during time Δt, an electromagnetic wave passes a particular point in space, then the length of the "volume" containing the field is $l = c\Delta t$. If we consider a $1m^2$ area *perpendicular* to the direction of travel, then the intensity of the electromagnetic field is obtained from:

$$I = \frac{P}{Area}$$

$$= \frac{u}{\Delta t} V \frac{1}{A}$$

$$= \frac{u}{\Delta t} lA \frac{1}{A}$$

$$= \frac{u}{\Delta t} c\Delta t$$

$$= uc$$

The use of the perpendicular area is a hint: a since E and B are vectors, and it is the perpendicular area through which the intensity is being calculated, the intensity I will be a vector also and will involve a cross product.

Substituting the energy density u, we have:

$$I = \varepsilon_0 E^2 c$$

$$= \varepsilon_0 EB \frac{1}{\varepsilon_0 \mu_0} \quad \text{since}$$

$$= \frac{1}{\mu_0} EB \qquad c = \frac{1}{\sqrt{\mu_0 \varepsilon_0}}$$

In terms of **magnetic field intensity** H, and taking into account the vector nature of E and H, the intensity I is expressed:

$$\boxed{\overline{S} = \overline{E} \times \overline{H}}$$

The vector S is the **Poynting vector**.

Average intensity of electromagnetic wave

If E and B both vary sinusoidally with time, and E_0 and B_0 give a maximum intensity I_0, then:

$$I = \frac{1}{\mu_0} E_0 B_0 \sin^2(\omega t)$$

To find the average power, or average intensity, we integrate the $\sin^2$ function over 2π::

$$I_{av} = \frac{\displaystyle\int_0^{2\pi} I_i d\theta}{2\pi}$$

this integral evaluates to π

$$= \frac{1}{\mu_0} \frac{E_0 B_0}{2\pi} \int_0^{2\pi} \sin^2 \theta d\theta$$

$$= \frac{1}{2\mu_0} E_0 B_0$$

Or, working in terms of the E field only:

$$I_{av} = \varepsilon_0 c \left[\frac{1}{2} E_0^{\,2} \right] \longrightarrow E_{av}^{\,2} = \frac{1}{2} E_0^{\,2}$$

$$= \varepsilon_0 c E_{av}^{\,2} \qquad \text{rms value of E}$$

3.7.11 Examples

1. Calculate the magnitude of the magnetic field in an electromagnetic wave if the magnitude of the electric field is measured to be 10 V m^{-1}.

Solution:

$$B = \frac{E}{c}$$

$$= \frac{10}{3 \times 10^8}$$

$$= 3.33 \times 10^{-8} \text{ T}$$

2. The radiation output from the sun is estimated to be 4×10^{23} kW. (a) Calculate the radiation intensity at the edge of the atmosphere if the distance from the sun is taken to be 1.49×10^{11} m. (b) Determine the average magnitude of the E and B fields of this radiation. Compare B with the magnitude of the Earth's magnetic field B_E

Solution:

(a)

$$I_{av} = \frac{P}{A}$$

$$= \frac{4 \times 10^{23}}{4\pi \left(1.49 \times 10^{11}\right)^2}$$

$$= 1.4 \text{ k W m}^{-2}$$

(b)

$$I_{av} = \varepsilon_o c E_{av}{}^2$$

$$1.4 \times 10^3 = \left(8.85 \times 10^{-12}\right)\left(3 \times 10^8\right) E_{av}{}^2$$

$$E_{av} = 726 \text{ V m}^{-1}$$

$$B_{av} = \frac{726}{3 \times 10^8}$$

$$= 2.42 \times 10^{-6} \text{ T}$$

$$B_E \approx 10^{-4} \text{ T}$$

Earth's magnetic field is two orders of magnitude larger.

Part 4

Mechanics

4.1 Scalars and vectors

Summary

$$|R| = \sqrt{R_H^2 + R_V^2}$$ Magnitude of vector

$$\tan \theta = \frac{|R_V|}{|R_H|}$$ Angle between vectors

$$\mathbf{R} = A\mathbf{i}, B\mathbf{j}, C\mathbf{k}$$ Vector components

$$|R| = \sqrt{A^2 + B^2 + C^2}$$ Magnitude of resultant

$$\mathbf{A} = A_1\mathbf{i} + B_1\mathbf{j} + C_1\mathbf{k}$$ Unit vector

$$\mathbf{B} = A_2\mathbf{i} + B_2\mathbf{j} + C_2\mathbf{k}$$

$$\mathbf{A} \bullet \mathbf{B} = |A||B|\cos\theta$$ Vector dot product

$$\mathbf{A} \bullet \mathbf{B} = A_1A_2 + B_1B_2 + C_1C_2$$

$$\mathbf{C} = \mathbf{A} \times \mathbf{B}$$ Vector cross product

$$|C| = |\mathbf{A}||\mathbf{B}|\sin\theta$$

4.1.1 Vectors

Physical quantities that have a direction associated with them are called **vectors**.

Examples are: displacement, velocity, acceleration, angular acceleration, torque, momentum, force, weight.

Physical quantities that do not have a direction associated with them are called **scalars**.

Examples are: temperature, energy, mass, electric charge, distance.

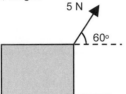

A good example of a **vector quantity** is a **force**. Let's look at a force of 5 N pulling upwards on a body at an angle of 60°. The vector that represents this force consists of a **magnitude** and a **direction.**

We draw an arrow at θ = 60° and length 5 units on the x and y coordinate axes to represent the force acting on the body.

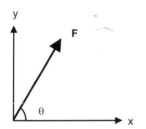

Can we replace this force with two forces F_X and F_Y which act in the vertical and horizontal directions, and which when acting together at the same time, produce the exact same motion of the body as the single 5N force acting at 60°?

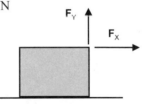

The answer is of course, yes! All we do is lay out on the x and y axes arrows which correspond to the force at θ = 60°.

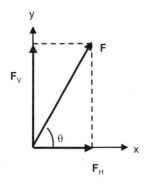

These horizontal and vertical forces, which when acting together have the same effect as the original force, are called the **force components** of **F**. We can call these components F_H and F_V, or F_X and F_Y, it doesn't matter, as long as we know which component is which. The magnitude of these components are simply:

$$|F_H| = |F|\cos\theta$$
$$|F_V| = |F|\sin\theta$$

4.1.2 Addition of vectors

Dividing up a vector into components is a useful procedure because it allows us to easily calculate the effect of a combination of vectors.

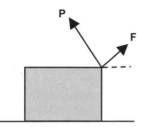

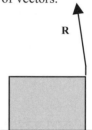

The combined effect of the two forces can be expressed as a single force which we call the **resultant R**.

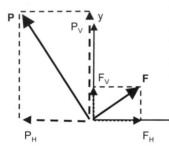

To find the resultant of **F** and **P**, we simply add the vertical components of each together to obtain the vertical component of the resultant, and do the same for the horizontal components.

$$R_V = F_V + P_V$$

R$_V$ and **R**$_H$ are the components of the resultant force **R**.

$$R_H = F_H + P_H$$

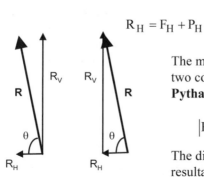

The magnitude of resultant R of these two components is given by **Pythagoras' theorem**:

$$|R| = \sqrt{R_H^2 + R_V^2}$$

The direction, or angle, of the resultant is found from:

$$\tan\theta = \frac{|R_V|}{|R_H|}$$

There is a graphical method of determining the magnitude and direction of vectors. We draw vectors in a head to tail manner. The resultant R is found by joining up any gap between the first tail and the last head.

4.1.3 The unit vectors

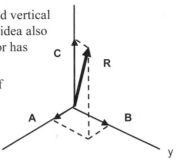

The division of a vector into horizontal and vertical components is a very useful concept. The idea also applies in three dimensions, where a vector has components along the x, y and z axes.

That is, the direction of the components of the vector are that of the corresponding coordinate axis. Let the magnitude of these components be A, B and C.

Rather than saying a vector **R** has a magnitude of 5 units and acts at 30° to the x axis, 20° to the y axis, and 80° to the z axis, we need a good concise method of expressing this information.

Let the vectors **i**, **j** and **k** have a magnitude of 1 unit, and have directions along the x, y and z axes respectively. Now, the components of our vector can be written:

$$\mathbf{R} = A\mathbf{i}, B\mathbf{j}, C\mathbf{k}$$

Why do this? It is a good way to keep the components of a vector organised. Indeed a great way of writing a vector in terms of its components is: $\mathbf{R} = A\mathbf{i} + B\mathbf{j} + C\mathbf{k}$

Magnitude ← ┘ └→ Direction

This is a **vector equation**. We express a vector as the resultant of its component vectors. The magnitude of the component vectors are the scalars A, B and C. When the unit vector **i** (which points in the x direction) is multiplied by the scalar magnitude A, we obtain A**i** which is the vector component in the x direction and so on for the product B**j** and C**k**. The magnitude of **R** is given by:

$$|R| = \sqrt{A^2 + B^2 + C^2}$$

When two vectors are to be added, we simply add together the corresponding magnitudes of the **i** unit vector, the **j** and the **k** unit vectors.

$$\mathbf{A} = A_1\mathbf{i} + B_1\mathbf{j} + C_1\mathbf{k}$$

$$\mathbf{B} = A_2\mathbf{i} + B_2\mathbf{j} + C_2\mathbf{k}$$

$$\mathbf{A} + \mathbf{B} = A_1\mathbf{i} + B_1\mathbf{j} + C_1\mathbf{k} + A_2\mathbf{i} + B_2\mathbf{j} + C_2\mathbf{k}$$
$$= (A_1 + A_2)\mathbf{i} + (B_1 + B_2)\mathbf{j} + (C_1 + C_2)\mathbf{k}$$

4.1.4 Vector dot product

When force is multiplied by distance, we obtain the **work** done by the force. But the force and the distance must be acting in the same direction.

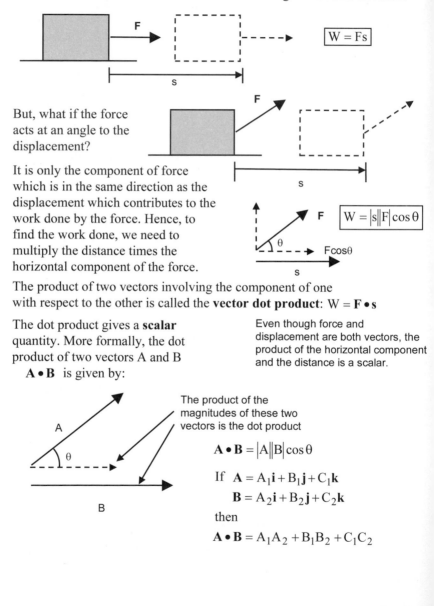

But, what if the force acts at an angle to the displacement?

It is only the component of force which is in the same direction as the displacement which contributes to the work done by the force. Hence, to find the work done, we need to multiply the distance times the horizontal component of the force.

The product of two vectors involving the component of one with respect to the other is called the **vector dot product**: $W = \mathbf{F} \bullet \mathbf{s}$

The dot product gives a **scalar** quantity. More formally, the dot product of two vectors A and B $\mathbf{A} \bullet \mathbf{B}$ is given by:

Even though force and displacement are both vectors, the product of the horizontal component and the distance is a scalar.

The product of the magnitudes of these two vectors is the dot product

$$\mathbf{A} \bullet \mathbf{B} = |A||B|\cos\theta$$

If $\mathbf{A} = A_1\mathbf{i} + B_1\mathbf{j} + C_1\mathbf{k}$
  $\mathbf{B} = A_2\mathbf{i} + B_2\mathbf{j} + C_2\mathbf{k}$

then

$$\mathbf{A} \bullet \mathbf{B} = A_1A_2 + B_1B_2 + C_1C_2$$

4.1.5 Vector cross product

When a force is applied to a body so as to cause a rotation, we say that the application of the force results in there being a **moment** applied about the axis of rotation.

The magnitude of the moment is given by the product of the force times the distance from the axis of rotation.

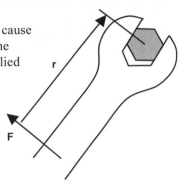

$$\left|T\right| = \left|F\right|\left|r\right|$$

The moment, or **torque** only depends on the perpendicular distance between the axis of rotation and the line of action of the force. Thus, if the force is applied at an angle, the moment is given by the vertical component of the force times the distance.

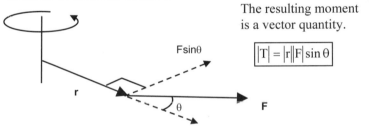

The resulting moment is a vector quantity.

$$\left|T\right| = \left|r\right|\left|F\right|\sin\theta$$

The product of the vertical component and the perpendicular distance yields a new vector perpendicular to both.

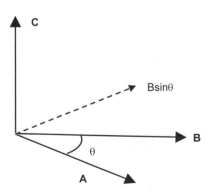

The direction is given by the right hand rule. If the fingers curl in the direction from **A** to **B**, then the direction of the resulting vector **AxB** is given by the thumb. Thus, **AxB** is opposite in direction to **BxA**.

$$C = A \times B$$
$$\left|C\right| = \left|A\right|\left|B\right|\sin\theta$$

4.1.6 Scalars and vectors (summary)

Quantities like displacement, velocity, acceleration, angular acceleration, torque, momentum, force, all have both a magnitude and a direction associated with them, they are **vector quantities**.

Quantities such as temperature, energy, mass, electric charge do not have a direction associated with them. For example, a mass moving with a velocity has the same amount of kinetic energy no matter what the direction of the velocity might be. They are **scalar quantities**.

A **vector dot product** gives a scalar quantity.

$$\mathbf{A} \bullet \mathbf{B} = |A||B| \cos \theta$$

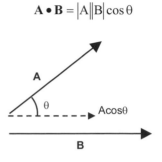

A **vector cross product** gives a vector quantity.

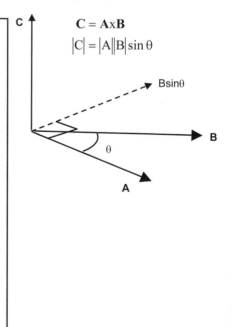

$$C = A \times B$$
$$|C| = |A||B| \sin \theta$$

Dot and cross products

Why does the dot product give a scalar quantity and the cross product a vector quantity? This is a hard question to answer. The dot product A.B gives the same result no matter which coordinate axes are used to represent the two vectors A and B. For the cross product, the two magnitudes of the two vector components being multiplied are at right angles to each other. The resulting product is a number, which in two dimensional space, would be a scalar. But, in three dimensions, the cross product behaves mathematically like a vector. Mathematically, it appears that this new vector points in a direction normal to the plane formed by the other two vectors. That is, the vector nature of the cross product is a consequence of the three dimensional character of our space and laws of mathematics!

4.1.7 Example

1. Find the difference A-B between the two vectors shown
    below:

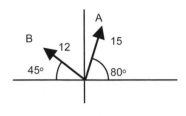

Solution:

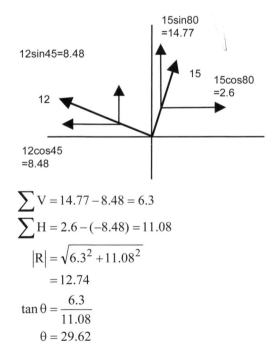

$$\sum V = 14.77 - 8.48 = 6.3$$
$$\sum H = 2.6 - (-8.48) = 11.08$$
$$|R| = \sqrt{6.3^2 + 11.08^2}$$
$$= 12.74$$
$$\tan \theta = \frac{6.3}{11.08}$$
$$\theta = 29.62$$

4.2 Statics

Summary

$$\sum F_H = 0$$ Static equilibrium

$$\sum F_V = 0$$

$$\sum M = 0$$

4.2.1 Equilibrium

Equilibrium is a state of balance.
When an object (or **body**) is at rest,
then we say that the body is in **static
equilibrium** and the vector sum of all
forces acting on a body is zero..

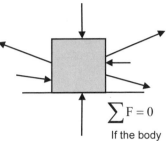

$$\sum F = 0$$

If the body
is at rest

When a body is moving with a
constant velocity, then the body
is in **dynamic equilibrium**.

To make it easy to determine the vector sum of the
forces, it is convenient to divide them up into
horizontal and vertical components.

$$\sum F_H = 0$$
$$\sum F_V = 0$$

This is the **Force
Law of Equilibrium**

The force law of equilibrium is useful for
determining the magnitude and directions of
unknown forces on a body at rest.

Static equilibrium also means that the body is not
rotating. Hence, for static equilibrium, the sum of the
moments acting on the body must equal zero.

$$\sum M = 0$$

When determining moments acting on a body, we must choose clockwise or anti-
clockwise as being positive and stick to it. It doesn't matter which direction is chosen,
some text books say clockwise is positive, others anti-clockwise.

When two forces are in
equilibrium, they must be equal
and opposite in direction.

When three forces are in
equilibrium, their lines of action
intersect at a common point (they
are **concurrent**) and act in the
same plane (they are **coplanar**).

When four forces are in equilibrium,
the resultant of any two of them
must be equal and opposite to the
resultant of the other two.

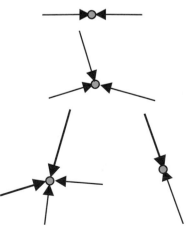

4.2.2 The normal force

Newton's 3rd law states that every action is accompanied by an equal and opposite reaction.

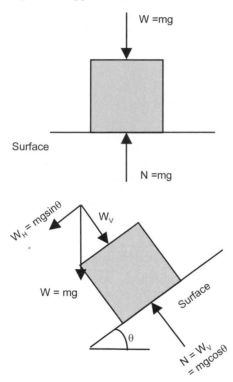

When an object rests on a surface, the **gravitational force** W acts downwards. The reaction to this downwards force is what we call the **normal** force acting upwards. The normal force N is equal and opposite to W.

On an **inclined surface**, the normal force has a magnitude equal and opposite to that of the vertical component of the weight W.

When the object rests on the surface, the atoms within the surface actually deflect from their equilibrium positions in the crystal structure of the material. This deflection is much like that of a deflected spring. The deflection results in their being a resisting force much like that when we push our hands down and compress a spring. Every surface deflects a very small amount when placed under load. This deflection results in the reaction force.

4.2.3 Reactions

The normal force is one example of a reaction to an imposed load. Consider a **truss** of a bridge which has a **pin joint** at one end and a **sliding joint** on the other.

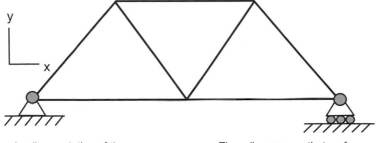

The pin allows rotation of the truss but prevents linear movement in the x-y directions.

The rollers ensure that no forces are transmitted in a direction parallel to the supporting surface and allows the truss to expand or contract according to a change in load or temperature

Now, the truss can be considered to be a rigid body (comprised of individual members). In two dimensions, the laws of equilibrium can only be used if the body is supported by no more than two reaction forces. If there are more than two (or three in three dimensions), then the reaction forces are **indeterminate** since it is generally not possible to determine what proportion of the load is carried by each support. Thus, in the case of the truss, we need to show that the pin and roller supports correspond to two and only two effective support or reaction forces.

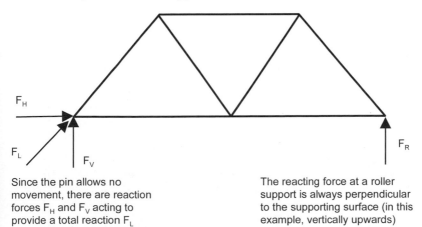

Since the pin allows no movement, there are reaction forces F_H and F_V acting to provide a total reaction F_L

The reacting force at a roller support is always perpendicular to the supporting surface (in this example, vertically upwards)

4.2.4 Free body diagram

Earlier, we showed how the supports offered to a body subjected to **gravitational** and other loads can be shown as reaction forces. This is the procedure for drawing a **free-body diagram**.

> A diagram of a body which is isolated from all its supports and shows only those forces acting directly on it is called a free body diagram.

It so happens that part of an object may be represented as a free body diagram as long as the internal forces acting on the isolated member from the part removed are shown.

This is important

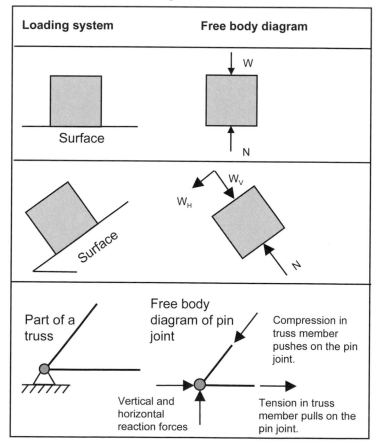

Loading system	Free body diagram

Surface

W

N

W_V

W_H

Surface

N

Part of a truss

Free body diagram of pin joint

Compression in truss member pushes on the pin joint.

Vertical and horizontal reaction forces

Tension in truss member pulls on the pin joint.

4.2.5 Resultants

The equations of equilibrium can be used to find
the **resultant** of a force system. Consider the
forces on the beam shown below:

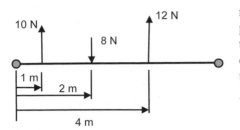

Can all these forces be
replaced by a single force
positioned in such a way so
that the reaction forces at each
end of the beam are
unchanged?

Yes!

To do this, we imagine that the resultant is the sum of the individual forces.
In this case, all the forces are vertical, and thus, taking upwards to be
positive, we can see that the magnitude of the resultant is:

$$R = 10 + 12 - 8$$
$$= 14\text{N}$$

Upwards.

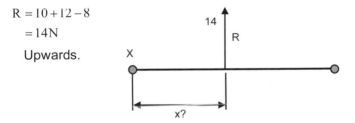

But where should this 14 N upwards force be located? We take
moments about a point X. Taking anticlockwise as positive, we have:

$$14x = 10(1) + -8(2) + 12(4)$$
$$x = 3\,\text{m}$$

Note, an anticlockwise moment is taken
as positive since here, we chose
upwards as a positive direction for
forces, and to the right, as a positive
direction for distance.

We don't show the reaction forces at the end of the bar since we want the
resultant of the applied forces. If the reaction forces were included, the
resultant would be zero because the bar is in static equilibrium.

4.2.6 Example

1. A ladder leans against a smooth wall.
   George, who has a mass of 81.5 kg,
   has climbed up the ladder and is
   resting at a point C as shown. His
   friend Fred pushes against the ladder
   at D with a force 240 N to steady it.
   Calculate the force exerted on the
   wall and on the floor by the ladder.

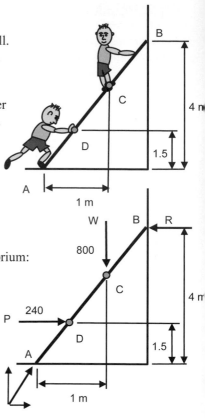

Solution:

The first step in solving this problem,
and all mechanics problems, is to <u>draw
a free-body diagram</u>.

Now, start with the equations of equilibrium:

$$\sum F_H = 0$$
$$-R + 240 + F_H = 0$$
$$F_H = R - 240$$

Can't do any more than this, so go
on to verticals.

$$\sum F_V = 0$$
$$800 = F_V$$

Well, this is fine for the vertical reaction at
A, but we still need some more information
to determine R. Let's try summing the
moments, say about "A".

$$\sum M_A = 0$$
$$240(1.5) + 800(1) = 4R$$
$$R = 290$$
$$F_H = 290 - 240$$
$$= 50N$$

4.3 Moment of inertia

Summary

$$\overline{x} = \frac{\sum ax}{A}$$
Centroid of an area

$$\overline{y} = \frac{\sum ay}{A}$$

$$AX^2 = \sum ax^2$$
$$= I_Y$$
2nd moment of an area or moment of inertia

$$AY^2 = \sum ay^2$$
$$= I_X$$

$$I_X = I_{XC} + Ad^2$$
Transfer formula

$$k = \sqrt{\frac{I}{A}}$$
Radius of gyration

4.3.1 Centroid and centre of mass

Gravity acts on all the atoms in a body, but it is convenient (for the purposes of mechanical analysis) to imagine that the gravitational force acts at a single point. The **centre of mass** is that point in the body through which the force of gravity can be said to act. The position of the centre of mass for an object can be calculated by considering its dimensions. Even complicated shaped bodies can be readily analysed.

To do this, let us imagine that a body is reduced in thickness to that of a sheet of paper. The problem is thus reduced to two dimensions. The point which used to be the centre of mass is now called the **centroid** of the area.

Centroid

The two dimensional surface area can be considered to be composed of a series of small elements, each of an area "a" and thickness "t" and density ρ.

The weight of one small element is thus:

$$w = \rho t a$$

The total weight is:

$$W = \sum w$$
$$= \sum \rho t a g$$
$$= \rho t A g$$

Now, the centroid of an area is the centre of mass for a thin object. If the object is supported at its centre of mass, it is in **static equilibrium**.

Total area.

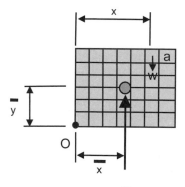

Let us take moments about the point O. For equilibrium:

$$W\bar{x} = w_1 x_1 + w_2 x_2 + w_3 x_3 ...$$
$$= \sum wx$$
$$\rho t A \bar{x} = \rho t \sum ax$$
$$\bar{x} = \frac{\sum ax}{A}$$

These equations give the coordinates of the centroid of the area A. We have neglected the factor g for convenience here.

If the object were tipped up on its side, then:

$$\bar{y} = \frac{\sum ay}{A}$$

4.3.2 Centroids of areas

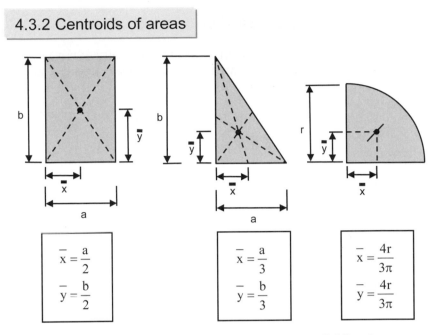

$$\bar{x} = \frac{a}{2}$$

$$\bar{y} = \frac{b}{2}$$

$$\bar{x} = \frac{a}{3}$$

$$\bar{y} = \frac{b}{3}$$

$$\bar{x} = \frac{4r}{3\pi}$$

$$\bar{y} = \frac{4r}{3\pi}$$

The **centroid** of a **composite area** can be calculated by dividing the composite area into simple areas and taking the moment about a convenient fixed point. We then treat each simple area as an element where:

$$\bar{x}A = \sum ax$$

$$\bar{y}A = \sum ay$$

The **moment of an area** is equal to the sum of the moments of its component areas.

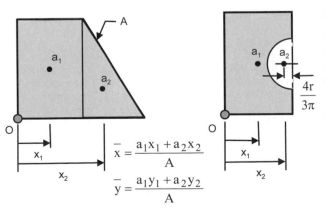

$$\bar{x} = \frac{a_1 x_1 + a_2 x_2}{A}$$

$$\bar{y} = \frac{a_1 y_1 + a_2 y_2}{A}$$

The moment of an area with area removed is the moment of the whole area minus that of the area removed.

$$\bar{x} = \frac{a_1 x_1 - a_2 x_2}{A}$$

$$\bar{y} = \frac{a_1 y_1 - a_2 y_2}{A}$$

4.3.3 2nd moment of an area

A very important quantity for determining the stress carrying capacity of beam is the **2nd moment of an area** - often called the **moment of inertia of an area** and given the symbol I.

The 2nd moment of an area is determined in a similar way to the first moment of an area except the distances are squared.

$$AX^2 = \sum ax^2$$
$$= I_Y$$

This is the moment of inertia w.r.t. the y axis.

The distances X and Y are not the centroid but represent locations called the radius of gyration.

For beams, a large moment of inertia about a particular axis is a measure of its load carrying capacity. For example, an **I section beam** has a larger moment of inertia than a simple square section of the same cross sectional area because its mass is concentrated at larger distance away from the axis.

$$AY^2 = \sum ay^2$$
$$= I_X$$

This is the moment of inertia w.r.t. the x axis.

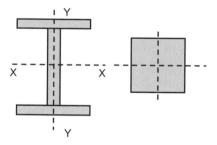

The distance X (or Y) is the **radius of gyration** about the y (or x) axis and is usually given the symbol k. The radius of gyration is the perpendicular distance from the axis at which the total mass or area of the body may be concentrated without changing its moment of inertia.

$$I_X = Ak^2$$

$$k = \sqrt{\frac{I}{A}}$$

Moments of inertia and moments of areas
In this chapter, we have talked about moments of areas, and moments of inertia of areas. We must not get these confused with the moment of inertia of a rotating body. The two moments of inertia are similar, but one applies to an <u>area</u>, the other to a <u>body</u>. Thus when talking about moments of inertia, we must specify what we are talking about. Hence, unless the context is clear, we say "moment of inertia of an area" or "moment of inertia of a body" about a particular axis.

Note, since we are taking about moments of inertia, (i.e the *second* moment of an area), the radius of gyration does not correspond to the coordinates of the centroid (the first moment of an area) about an axis since the second moment of an area contains a a sum of distance *squared* terms.

4.3.4 Moments of inertia of areas

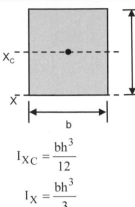

$$I_{XC} = \frac{bh^3}{12}$$

$$I_X = \frac{bh^3}{3}$$

$$I_{XC} = \frac{bh^3}{36}$$

$$I_X = \frac{bh^3}{12}$$

$$I_{XC} = \frac{\pi r^4}{4}$$

The units of 2nd moment of area or moment of inertia of an area are m⁴

The **moment of inertia** of a **composite area** can be determined using the moments of inertia of the component simple areas.

$$I_{XC} = \left(\frac{\pi}{8} - \frac{8}{9\pi} \right) r^4$$

$$I_X = \frac{\pi r^4}{8} = I_{YC}$$

- Determine the moment of inertia of the area component with respect to a parallel **centroidal axis**.
- Add the product of the area component and the distance from the centroidal axis to the axis XX squared.

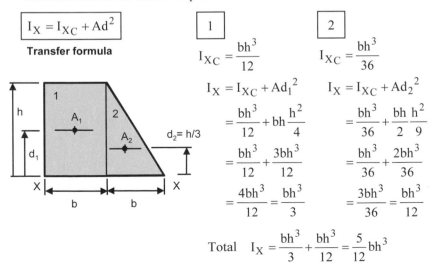

$$\boxed{I_X = I_{XC} + Ad^2}$$

Transfer formula

1

$$I_{XC} = \frac{bh^3}{12}$$

$$I_X = I_{XC} + Ad_1^2$$

$$= \frac{bh^3}{12} + bh \frac{h^2}{4}$$

$$= \frac{bh^3}{12} + \frac{3bh^3}{12}$$

$$= \frac{4bh^3}{12} = \frac{bh^3}{3}$$

2

$$I_{XC} = \frac{bh^3}{36}$$

$$I_X = I_{XC} + Ad_2^2$$

$$= \frac{bh^3}{36} + \frac{bh}{2} \frac{h^2}{9}$$

$$= \frac{bh^3}{36} + \frac{2bh^3}{36}$$

$$= \frac{3bh^3}{36} = \frac{bh^3}{12}$$

Total $I_X = \frac{bh^3}{3} + \frac{bh^3}{12} = \frac{5}{12} bh^3$

4.3.5 Example

1. Determine the moment of inertia of
   the cross-sectional area shown:

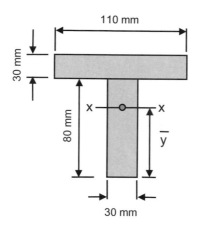

Solution:

$$\bar{y} = \frac{(0.110)(0.03)(0.095) + (0.08)(0.03)(0.04)}{0.110(0.03) + 0.08(0.03)}$$

Find location of centroid
of the area.

$$= 0.072 \text{m}$$

$$I_{xx1} = \frac{(0.110)(0.03^3)}{12} + (0.11)(0.03)(0.095 - 0.072)^2$$

$$= 1.99 \times 10^{-6} \text{m}^4$$

$$I_{xx2} = \frac{(0.03)(0.08^3)}{12} + (0.08)(0.03)(0.072 - 0.040)^2$$

$$= 3.74 \times 10^{-6} \text{m}^4$$

$$I_{xx} = 5.73 \times 10^{-6} \text{m}^4$$

4.4 Linear motion

Summary

$$v = \frac{ds}{dt}$$ Velocity

$$v_{ac} = v_{ab} + v_{bc}$$ Relative motion

$$a = \frac{dv}{dt}$$ Acceleration

$$v = u + at$$ Kinematics

$$v^2 = u^2 + 2a\Delta s$$

$$\Delta s = ut + \frac{1}{2}a\Delta t^2$$

$$t_f = -\frac{2v_o \sin \theta}{g}$$ Time of flight

4.4.1 Velocity and acceleration

Velocity is a vector quantity that is a measure of a body's displacement per second. The term speed is used as a common name for the magnitude of a body's velocity. Bodies do not always move with a constant velocity. The **average velocity** is found from the total distance travelled over a time period.

$$v_{av} = \frac{\Delta s}{\Delta t}$$

The smaller the time interval selected, the more representative is the calculation of the **instantaneous velocity** at a particular time t.

$$v = \lim_{\Delta t \to 0} \frac{\Delta s}{\Delta t} = \frac{ds}{dt}$$

Acceleration is a measure of the magnitude of the change in velocity over a time period - that is, the rate of change of velocity with time.

The rate of change of velocity may not necessarily be constant. That is, the acceleration of a body may vary with time. If the time period is made small enough, then the **instantaneous acceleration** is

When the velocity increases in time, it is called an **acceleration**. When the velocity decreases with time, it is called a **deceleration**.

$$a = \lim_{\Delta t \to 0} \frac{\Delta v}{\Delta t} = \frac{dv}{dt}$$

Relative motion

It is often convenient to describe the motion of one body in terms of the motion of another.

This is easily done because velocity is a vector, all we need do is add the components. The general equation is:

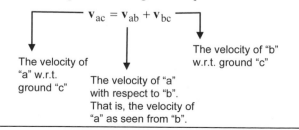

$$\mathbf{v}_{ac} = \mathbf{v}_{ab} + \mathbf{v}_{bc}$$

The velocity of "a" w.r.t. ground "c"

The velocity of "a" with respect to "b". That is, the velocity of "a" as seen from "b".

The velocity of "b" w.r.t. ground "c"

This is a vector equation - cannot just add the magnitudes of the velocities.

4.4.2 Kinematics

Consider the mathematical definitions of velocity and acceleration:

$$v = \frac{ds}{dt} \quad \text{velocity} \qquad a = \frac{dv}{dt} \quad \text{acceleration}$$

$$dt = \frac{ds}{v} \qquad\qquad = \frac{dv}{ds} v \qquad \text{Substituting dt = ds/v}$$

$$\boxed{a\,ds = v\,dv}$$ This is called a **differential equation** because it contains **differentials** (ds and dv).

Let us assume acceleration, a, is a constant and at time $t = 0$, the displacement $s = s_0$ and the velocity $v = u$, the **initial velocity**.

1.
$$a = \frac{dv}{dt}$$
$$dv = a\,dt$$
$$v = \int a\,dt$$
$$= at + C$$
$$@\,t = 0, v = u$$
$$\therefore C = u$$
$$\boxed{v = u + at}$$

2. $$a\,ds = v\,dv$$
$$\int_{s_0}^{s} a\,ds = \int_{u}^{v} v\,dv$$
$$a(s - s_0) = \left[\frac{1}{2}v^2\right]_{u}^{v}$$
$$= \frac{1}{2}\left[v^2 - u^2\right]$$
$$2a(s - s_0) = v^2 - u^2$$
$$\boxed{v^2 = u^2 + 2a\Delta s}$$

3.
$$v = u + at$$
$$= \frac{ds}{dt}$$
$$(u + at)dt = ds$$
$$\int_{0}^{t}(u + at)dt = \int_{s_0}^{s} ds$$
$$ut + \frac{1}{2}at^2 = s - s_0$$
$$\boxed{\Delta s = ut + \frac{1}{2}a\Delta t^2}$$

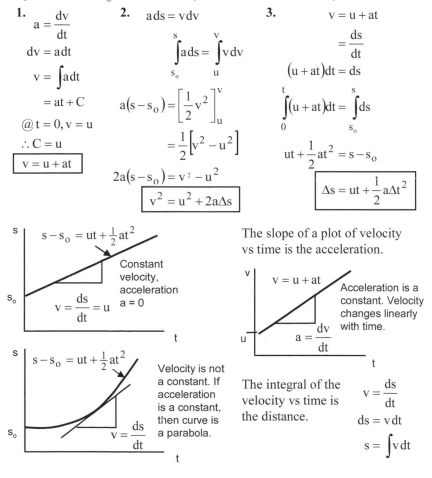

$$s - s_0 = ut + \tfrac{1}{2}at^2$$

Constant velocity, acceleration
$$v = \frac{ds}{dt} = u \qquad a = 0$$

$$s - s_0 = ut + \tfrac{1}{2}at^2$$

Velocity is not a constant. If acceleration is a constant, then curve is a parabola.
$$v = \frac{ds}{dt}$$

The slope of a plot of velocity vs time is the acceleration.

$$v = u + at$$

Acceleration is a constant. Velocity changes linearly with time.
$$a = \frac{dv}{dt}$$

The integral of the velocity vs time is the distance.

$$v = \frac{ds}{dt}$$
$$ds = v\,dt$$
$$s = \int v\,dt$$

4.4.3 Projectile motion

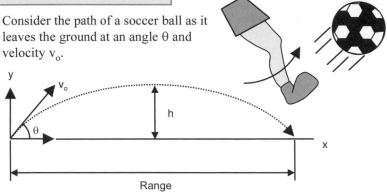

Consider the path of a soccer ball as it leaves the ground at an angle θ and velocity v_o.

The key to analysing **projectile motion** is to divide the motion into horizontal and vertical components.

Horizontal

$$v_{ox} = v_o \cos \theta$$

If we ignore air resistance and other losses, then v_x remains constant until the ball strikes the ground at the end of the travel. That is, the acceleration of the projectile in the x direction is zero.

From: $s = ut + \frac{1}{2}at^2$

with $u = v_{ox}$

$$x = (v_o \cos \theta)t$$

Note, when t = time of flight, x is the **range**.

Vertical

$$v_{oy} = v_o \sin \theta$$

The vertical component v_y changes at the rate of g = −9.81 m s^{-2} due to gravity. v_y does not remain constant during the flight of the ball. That is, the acceleration of the projectile in the y direction is −9.81 m s^{-2}.

$$y = (v_o \sin \theta)t + \frac{1}{2}gt^2$$

with the −9.81 indicated over the g term.

Note, when t = time of flight, the distance y (the distance above or below the **ground**) - is zero if starting and ending at ground level.

Now, the time of flight t_f depends only on the vertical motion.

$$y = (v_0 \sin \theta)t + \tfrac{1}{2}gt^2$$

When starting from ground level, $y = 0$, the time for the projectile to hit the ground occurs when $y = 0$ again. Hence:

$$(v_0 \sin \theta)t_f = -\tfrac{1}{2}gt_f^2$$

$$v_0 \sin \theta = -\tfrac{1}{2}gt_f$$

$$\boxed{t_f = -\frac{2v_0 \sin \theta}{g}}$$

When $g = -9.81$ is substituted,
t_f comes out positive

Note that the time of flight depends on the angle θ. This is important because the range depends on *both* θ and t.

$$x = (v_0 \cos \theta)t$$

Substituting t for time of flight thus gives the range:

$$x = (v_0 \cos \theta)\frac{-2v_0 \sin \theta}{g}$$

$$= \frac{-2v_0^2}{g}\sin \theta \cos \theta = \frac{-v_0^2 \sin 2\theta}{g}$$

The maximum range occurs when $dx/d\theta = 0$

$$\frac{dx}{d\theta} = \frac{-2v_0^2}{g}\left[-\sin^2 \theta + \cos^2 \theta\right]$$

$$= 0$$

$$\sin \theta = \cos \theta$$

$$\theta = 45°$$

The maximum height h occurs when $dy/dt = 0$:

i.e. Velocity in the vertical direction = 0

$$y = (v_0 \sin \theta)t + \tfrac{1}{2}gt^2$$

$$\frac{dy}{dt} = v_0 \sin \theta + gt$$

$$= 0$$

$$\boxed{t = -\frac{v_0 \sin \theta}{g}}$$

When $g = -9.81$ is substituted, t comes out positive

4.4.4 Examples

1. A bus leaves a bus stop after picking up passengers and accelerates at a rate of 2 m s^{-2} for 3 seconds, travels at constant velocity for 2 minutes, decelerates at a rate of 5 m s^{-2} and then comes to a stop at the next bus stop. Calculate the distance between the bus stops.

Solution:

(a) $a = 2$ m s^{-2}, $u = 0$, $t = 3$ s

$$s = ut + \tfrac{1}{2}at^2$$
$$= \tfrac{1}{2}2\!\left(3^2\right)$$
$$= 9\text{m}$$
$$v^2 = u^2 + 2as$$
$$= 0 + 2(2)9$$
$$v = 6\,\text{m s}^{-1}$$

(b) $a = 0$ ms^{-2}, $u = 6.0$ m s^{-1}, $v = 6.0$ m s^{-1}, $t = 2$ min

$$s = ut + \tfrac{1}{2}at^2$$
$$= 6(120)$$
$$= 720\,\text{m}$$

(c) $a = -5$ m s^{-2}, $u = 6.0$ m s^{-1}, $v = 0$

$$v^2 = u^2 + 2as$$
$$0 = 6^2 + 2(-5)s$$
$$s = 3.6\,\text{m}$$

Total distance $= 9 + 720 + 3.6 = 832.6$ m

2. A boat travelling east at 20 km h^{-1} while another is going North at 10 km h^{-1}. Calculate the velocity of the first boat as seen by the second?

Solution:

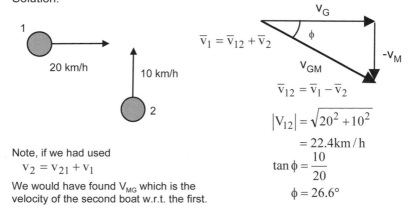

Note, if we had used
$$v_2 = v_{21} + v_1$$
We would have found V_{MG} which is the velocity of the second boat w.r.t. the first.

$$\bar{v}_1 = \bar{v}_{12} + \bar{v}_2$$
$$\bar{v}_{12} = \bar{v}_1 - \bar{v}_2$$
$$\left|V_{12}\right| = \sqrt{20^2 + 10^2}$$
$$= 22.4\,\text{km/h}$$
$$\tan\phi = \frac{10}{20}$$
$$\phi = 26.6°$$

4.5 Forces

Summary

$F = ma$	Newton's first law
$W = mg$	Weight force
$F = \mu N$	D'Alembert's principle
$\sum F_x = 0$	d'Alembert's principle
$\sum F_y = 0$	
$\sum M_x = 0$	Sliding or tipping

4.5.1 Newton's laws

So far we have talked about the motion of objects or **bodies**, that is, velocity and acceleration, without any consideration as to what is the cause of this motion. Accelerations of bodies arise due to the application of forces, specifically, **unbalanced forces**. The magnitude of the acceleration is proportional to the magnitude of the **unbalanced force**. The direction of the acceleration is in the same direction as the unbalanced force.

Consider a body at rest on a frictionless surface:

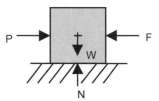

P balances F
N balances W
If all the forces on the body are balanced, then the body will remain at rest.

Let one of the applied forces P be removed.

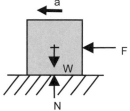

The forces on the body are not all balanced. In the vertical direction, W balances N. But in the horizontal direction, F is unbalanced.

The magnitude of the resulting acceleration is proportional to the magnitude of the force F and inversely proportional to the mass of the object.

The direction of the acceleration is in the same direction as the unbalanced force.

$$a = \frac{F}{m}$$

or $\boxed{F = ma}$

Consider a freely falling body of mass m. Neglecting air resistance, the only force acting on the body is the force of gravity. That is, the unbalanced force acting on the body is its weight W.

Experiment shows that the resulting acceleration, no matter what the mass of the body, is approximately 9.81m s^{-2}, the actual value depending on the location of the body on the earth. This value of the **gravitational acceleration** is given the symbol g.

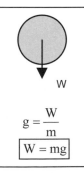

$$g = \frac{W}{m}$$

$\boxed{W = mg}$

4.5.2 Inertia

Consider a car towing a trailer. The car accelerates. Since the trailer is attached to the car by a draw bar, the trailer accelerates at the same rate as the car. From the point of view of someone standing on the footpath watching the trailer, the trailer is observed to experience an unbalanced force via the draw bar.

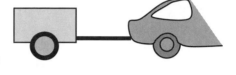

This unbalanced force causes the trailer to accelerate. The magnitude of the acceleration depends on the force applied to the draw bar by the car and the mass of the trailer.

Now, imagine you are driving the car which is towing the trailer and while accelerating, you look in the rear vision mirror. What do you see? A trailer behind you of course.

However, from your point of view in the accelerating car, you do not see the trailer accelerating or decelerating, the trailer is, from your point of view, motionless. It stays exactly in the same position at the rear of the car.

From your point of view, the trailer is in equilibrium, no unbalanced forces are acting on it, since if they were, it would be accelerating away from you. From the point of view of someone on the footpath, there *is* an unbalanced force, that in the draw bar, causing the trailer to accelerate.

From the point of view of you in the car, we say that the force in the draw bar acting on the trailer is balanced by an **inertia force**.

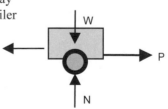

The inertia of a body *manifests itself as a force* when the body accelerates. The inertia force is always equal and opposite to the unbalanced force producing the acceleration.

The inertia force allows the motion of bodies being acted upon by forces to be analysed using equations of static equilibrium. This method of analysis is called **d'Alembert's principle**.

4.5.3 Ropes and pulleys

Consider the masses attached to a string
which passes over the pulleys as shown:

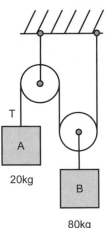

> In these systems of bodies, it is
> convenient to assume that the friction in
> the pulleys and the weight of the pulleys
> is negligible compared to the other
> forces involved and thus can be
> neglected. The tension T in the string is
> thus a constant throughout its length.

Careful consideration shows that the
displacement of A is twice that of B. To see
this, imagine that A moves downwards a
distance Δs. A point on the string on the left
side of pulley B moves upwards Δs.

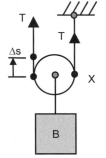

Pulley B rotates around a pivot point X and due
to the reduced leverage of the centre connection
to B, the mass B moves upwards $\Delta s/2$

Since the displacement of B is half that of A,
then the velocity and acceleration of B (if any)
will be half that of A.

In this system, B moves
downwards and A moves upwards.

To determine which way masses moves, we use the following systematic
approach. Let mass B be held stationary. The tension in the string is
produced by the weight of A, which in this case, is 20(9.81)=196.2 N. Now
let A be held stationary. The tension in the string is now provided by
weight B. But B is supported by 2T, hence, T is (80(9.81))/2=392.4 N.
Since the greatest tension in the string is produced when weight B pulls on
the system (with A stationary) then the direction of motion is that B moves
downwards.

The direction of motion is in the same direction as the direction of "pull"
by the body which is free to move with all the others held immobile which
produces the maximum tension.

4.5.4 Friction

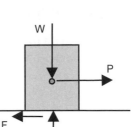

If two bodies are pressed together by a force normal to the contacting surfaces, then motion, or *attempted* motion, of one body with respect to the other in a direction parallel to the contact surface is resisted. This resisting force is called **friction**. In order to move one of the bodies, the maximum value of the frictional force must be overcome.

The maximum value of **friction force** is a measure of resistance to sliding of the two surfaces. For most contacting surfaces, the resistance to sliding depends on:

- The magnitude of the normal component of the force of contact N between the two surfaces
- The nature of the contacting materials
- The finish and state of the contacting surfaces
- Whether or not the surfaces are moving relative to one another

All these things can be grouped together and given a value called the **coefficient of friction**.

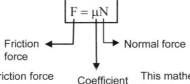

$$F = \mu N$$

Friction force ← | → Normal force

Note that the friction force does not depend on the area of contact between the bodies.

Coefficient of friction

This mathematical expression is called **Amontons' law** after Amontons who in 1699 performed many experiments on the frictional properties of materials.

Even the best-prepared surfaces are not entirely smooth. When two bodies are in contact, the asperities of each of them interlock to some extent. To move one of the bodies sideways, these asperities or protrusions must be sheared off. The coefficient of **static friction** is usually higher than that of **sliding friction** due to this initial shearing action on the microscopic scale.

In the case of static friction, Amonton's law gives the maximum value of friction force that can be achieved just prior to sliding. Below this force, the friction force is equal to the sideways force since the body remains at rest.

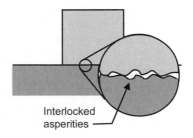

Interlocked asperities

4.5.5 Non-concurrent forces

The **inertia force** and d'Alembert's principle can be used to solve dynamics problems using equations of static equilibrium.

$$\sum F_x = 0$$

$$\sum F_y = 0$$

These equations are appropriate when all forces act through the centre of gravity - that is, the forces are **concurrent**.

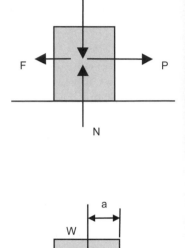

For many force systems, the actual points of application of forces are important and we must consider the equilibrium of moments as well as the forces:

$$\sum M_x = 0$$

$$(W)(a) = (P)(b)$$

In the case of a block resting on a flat surface, we have the force P acting on the centre of gravity, but the resulting friction force F acts along the surface. Both P and W produce moments around the point x. The forces F and N have no moment about x since the line of action of these forces passes through the point. When the moments about x are balanced, then the block is on the verge of tipping.

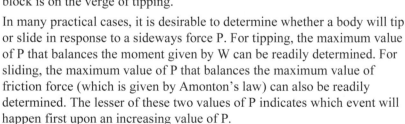

In many practical cases, it is desirable to determine whether a body will tip or slide in response to a sideways force P. For tipping, the maximum value of P that balances the moment given by W can be readily determined. For sliding, the maximum value of P that balances the maximum value of friction force (which is given by Amonton's law) can also be readily determined. The lesser of these two values of P indicates which event will happen first upon an increasing value of P.

4.5.6 Gravitation

In 1687, Newton proposed the law of universal gravitation. Application of this law was in accordance with the observed motions of the planets and **Kepler's laws**.

$$F = G \frac{m_1 m_2}{d^2}$$

The constant of proportionality, G, is called the **universal gravitational constant** and has the value $G = 6.673 \times 10^{-11}$ N m^2 kg^{-2}

The main idea of the law of universal gravitation is that every object in the universe attracts every other object. Precisely *why* this is so is not fully understood. The **gravitational force** is one of the four fundamental forces of nature, the others being the electrostatic force, and the weak and strong nuclear forces.

The gravitational force acts at a distance by some unknown mechanism. It is sometimes convenient to regard the effects of the gravitational force on a body in terms of a **gravitational field**. The motion of a particular object of mass m within a gravitational field of the Earth m_E at a distance d from the Earth can be readily calculated.

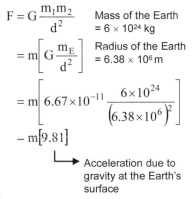

$$F = G \frac{m_1 m_2}{d^2}$$ Mass of the Earth = 6 × 10²⁴ kg

$$= m \left[G \frac{m_E}{d^2} \right]$$ Radius of the Earth = 6.38 × 10⁶ m

$$= m \left[6.67 \times 10^{-11} \frac{6 \times 10^{24}}{\left(6.38 \times 10^6 \right)^2} \right]$$

$$- m[9.81]$$

Acceleration due to gravity at the Earth's surface

Newton proved that when using the law of universal gravitation, the mass of a body can be thought of to exist at a single point at the centre of the body. This allows us to say that to calculate the gravitational force on an object on the Earth's surface, we assume that the whole mass of the Earth is concentrated at its centre and the distance d becomes the radius of the Earth.

The **Moon** is an average 381340 km from the Earth. The Moon is within the gravitational field of the Earth and so the force exerted on the Moon by the Earth is:

$$F = m \left[6.67 \times 10^{-11} \frac{6 \times 10^{24}}{\left(3.81 \times 10^8 \right)^2} \right]$$

$$= (m) 2.75 \times 10^{-3}$$

The Moon is falling towards the Earth with an acceleration of 2.75×10^{-3} m s^{-2}. Its orbital velocity keeps the Moon from actually falling into the Earth.

4.5.7 Examples

1. A person with mass 80 kg enters a lift on the 12th floor of a building and presses the ground floor button. Determine the force exerted by the person on the floor of the lift when:

   (a) The lift is stationary
   (b) The lift descends with an acceleration of 1 m s^{-2}
   (c) The lift comes to a stop with a deceleration of 1 m s^{-2}

> The only thing to remember about this method is that the inertia force F_I is:
>
> $$F_I = ma$$
>
> And acts in a direction opposite to that of the acceleration of the body.
> The equations of equilibrium are simply:
>
> $$\sum F_x = 0$$
>
> $$\sum F_y = 0$$

Solution:

(a) Draw a **free-body diagram** of the person:

Since the person is at rest, there is no inertia force.

$$N = W$$
$$= mg$$
$$= 80(9.81)$$
$$= 784.8N$$

A free-body diagram is a representation of an object or body which shows **all** forces acting upon it. The body is supported only by forces and is in contact with nothing else at all.

(b) Draw free-body diagram with inertia force directed opposite to acceleration.

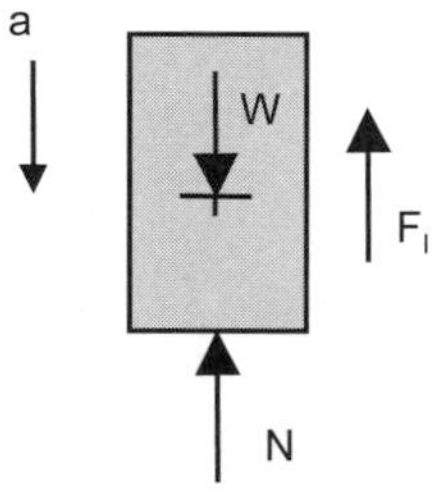

$$N + F_I = W$$
$$N + ma = mg$$
$$N = 80(9.81 - 1)$$
$$= 704.8N$$

(c) Draw free-body diagram with inertia force directed opposite to acceleration.

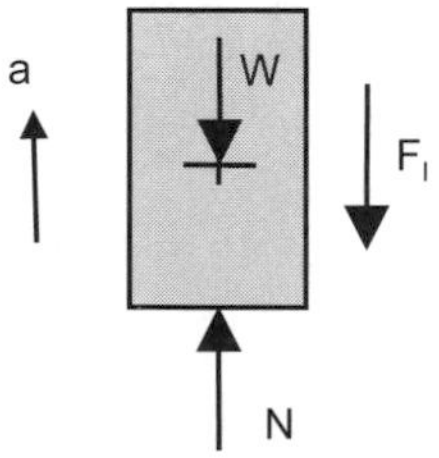

$$N = W + F_I$$
$$N = mg + ma$$
$$N = 80(9.81 + 1)$$
$$= 864.8N$$

2. A trailer has a mass of
    600 kg and has a rolling
    resistance of 700 N which
    represents rolling friction and
    air resistance. Calculate:

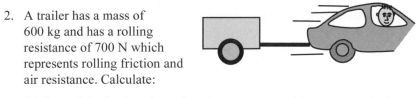

(a) Force P in the draw bar when the car moves with constant velocity

$$a = 0$$

R ← [ ] → P
$$P = R = 700N$$

(b) Force P in the draw bar when car accelerates at 1ms^{-2}

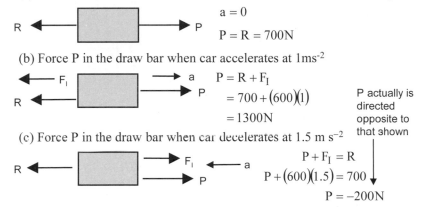

← F$_I$ [ ]   → a $P = R + F_I$
R ← [ ] → P $= 700 + (600)(1)$ P actually is
directed
$= 1300N$ opposite to
that shown

(c) Force P in the draw bar when car decelerates at 1.5 m s^{-2}

R ← [ ] → F$_I$ ← a $P + F_I = R$
            → P $P + (600)(1.5) = 700$ ↓
$$P = -200N$$

3. If the mass of the car is 800 kg, and has a rolling resistance of 900 N,
    calculate the force T required to be applied to the road by the wheels of
    the car for the accelerations and decelerations given previously.

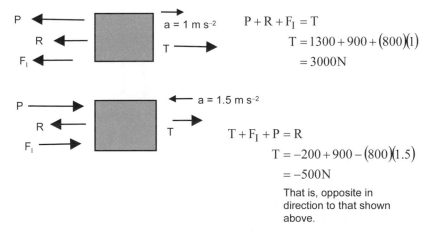

P ← [ ] → a = 1 m s^{-2} $P + R + F_I = T$
R ← [ ] → T $T = 1300 + 900 + (800)(1)$
F$_I$ ← $= 3000N$

P → [ ] ← a = 1.5 m s^{-2}
R ← [ ] → T $T + F_I + P = R$
F$_I$ → $T = -200 + 900 - (800)(1.5)$
$$= -500N$$

That is, opposite in
direction to that shown
above.

4. Calculate the mass of B required to give body A an acceleration of 2.4 m s^{-2} upwards.

Solution:

By inspection of the diagram, the magnitudes of displacement, velocity and acceleration of B are half that of A. Hence, we need body B to have a downwards acceleration of $2.4/2 = 1.2$ m s^{-2}. Now, considering the motion of A, we have

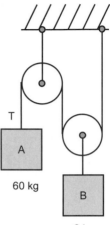

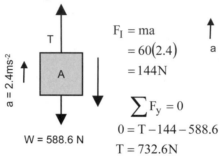

$F_I = ma$
$\quad = 60(2.4)$
$\quad = 144N$

$$\sum F_y = 0$$
$0 = T - 144 - 588.6$
$T = 732.6N$

Turning now to body B:

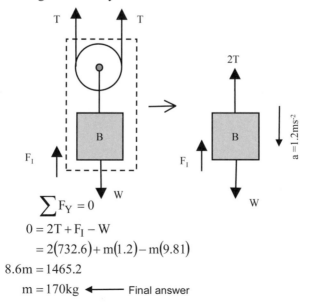

$$\sum F_Y = 0$$
$0 = 2T + F_I - W$
$\quad = 2(732.6) + m(1.2) - m(9.81)$
$8.6m = 1465.2$
$m = 170kg$ ◀───── Final answer

5. A box containing a new refrigerator is
    standing on the flat bed of a delivery
    truck. The block has a mass of 300 kg
    and is 1.2 m high and 0. 6 m wide.
    The coefficient of friction is 0.31. If
    the truck stops suddenly, will the box
    slide or tip?

Solution:

(a) Tipping   (b) Sliding

$$\sum M_a = 0$$

$$F_I(0.6) = 2943(.3)$$

$$F_I = 1471.5N$$

Inertia force has to be at
least this value for box to
be on the point of tipping.

$$F = \mu N = 0.31(2943)$$

$$= 912.3N$$

Inertia force has to be at least this value
before box will slide.

Since the inertia force will reach the value of friction force first as the
acceleration (and hence F_I) increases, then the box will slide and will not tip.

6. Calculate the gravitational force between two 1 kg masses placed
    1 cm apart.

Solution:

$$F = G\frac{m_1 m_2}{d^2}$$

$$= \left[6.673 \times 10^{-11}\frac{1}{(0.01)^2}\right]$$

$$= 6.673 \times 10^{-7}N$$

4.6 Rotational motion

Summary

$$\omega = \frac{d\theta}{dt} \qquad\qquad \text{Angular velocity}$$

$$\alpha = \frac{d^2\theta}{dt^2} = \frac{d\omega}{dt} \qquad \text{Angular acceleration}$$

$$\theta = \omega_o t + \frac{1}{2}\alpha t^2 \qquad \text{Equations of motion}$$

$$\omega = \omega_o + \alpha t$$

$$\omega^2 = \omega_o^2 + 2\alpha\theta$$

$$s = R\theta \qquad\qquad\quad \text{Linear displacement}$$

$$v_t = R\omega \qquad\qquad\quad \text{Tangential velocity}$$

$$a_t = R\alpha \qquad\qquad\quad \text{Tangential acceleration}$$

$$a_n = \frac{v_t^2}{R} \qquad\qquad\quad \text{Normal acceleration}$$

$$F_C = \frac{mv_t^2}{R} = m\omega^2 R \quad \text{Centripetal force}$$

4.6.1 Rotational motion

When a body moves in a straight line, this is called linear motion. Another common form of motion is when a body moves along the path of a circle. This is called **rotational motion**.

Consider a body moving in a circular path. Since the angular displacement is measured in terms of the angle (in **radians**) swept out by the body, then the rate of change of angular displacement with respect to time must be called the **angular velocity** and has the units rad s^{-1}.

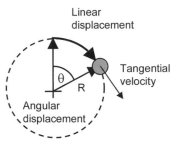

Linear displacement

Tangential velocity

Angular displacement

$$\omega = \frac{d\theta}{dt}$$

All points on a rotating (rigid) body have the same angular velocity.

The rate of change of angular velocity w.r.t. time is called **angular acceleration** and has the units: rad s^{-2} and is given the symbol α

$$\alpha = \frac{d^2\theta}{dt^2} = \frac{d\omega}{dt}$$

Equations of motion for rotational motion are very similar to those of linear motion:

Translational motion	Angular motion	
$s = ut + \dfrac{1}{2}at^2$	$\theta = \omega_0 t + \dfrac{1}{2}\alpha t^2$	Always best to convert to radians, radians per second, etc with these formulas.
$v = u + at$	$\omega = \omega_0 + \alpha t$	
$v^2 = u^2 + 2as$	$\omega^2 = \omega_0{}^2 + 2\alpha\theta$	

The **radian** is the angle swept out by an arc of length equal to the radius of the circle. The complete circumference of a circle can be calculated from $2\pi R$. The circumference is nothing more than an arc which sweeps through 360°. Thus, dividing the circumference by the radius gives us the number of radians in a complete circle.

$$2\pi \text{ radians} = 360°$$

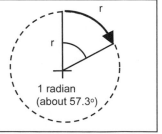

r

r

1 radian (about 57.3°)

4.6.2 Equations of motion

The radian is the angle which is swept out by an arc of length equal to the radius of the circle. Let the **arc length** be given the symbol s. Hence, the translational distance along the arc, for one radian, is equal to the radius R.

$$s = R\theta$$

The linear or **tangential velocity** and tangential acceleration of a point on the rotating body is found by differentiating:

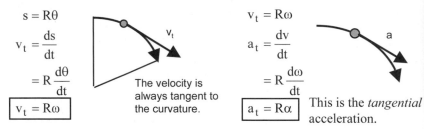

$$s = R\theta$$
$$v_t = \frac{ds}{dt}$$
$$= R\frac{d\theta}{dt}$$
$$\boxed{v_t = R\omega}$$

The velocity is always tangent to the curvature.

$$v_t = R\omega$$
$$a_t = \frac{dv}{dt}$$
$$= R\frac{d\omega}{dt}$$
$$\boxed{a_t = R\alpha}$$

This is the *tangential* acceleration.

Consider the motion of an object travelling in a circle with a constant **angular velocity**. For an **angular displacement** $\Delta\theta$, the **tangential velocity** changes *direction* from v_{t1} to v_{t2}.

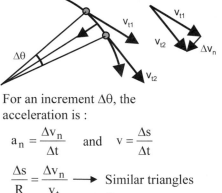

Δv_n represents the velocity difference and it is directed inwards towards the centre of rotation. This constant change in direction of the tangential velocity represents an **acceleration**, normal to the tangential velocity, and directed inwards towards the centre of rotation.

For an increment $\Delta\theta$, the acceleration is :

$$a_n = \frac{\Delta v_n}{\Delta t} \quad \text{and} \quad v = \frac{\Delta s}{\Delta t}$$

$$\frac{\Delta s}{R} = \frac{\Delta v_n}{v_t} \longrightarrow \text{Similar triangles}$$

$$a_n = \frac{\Delta v_n}{\Delta t}$$

$$= \frac{v_t \Delta s}{R\Delta t}$$

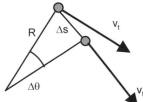

$$\boxed{a_n = \frac{v_t^2}{R}}$$

If the angular velocity of the object is changing, then the object accelerates in a tangential direction as well as in the normal direction.

4.6.3 Centripetal force

We have seen that motion of a body in
a curved or circular arc results in an
acceleration which is directed inwards
towards the **centre of rotation**, even
when the body has a constant
tangential speed.

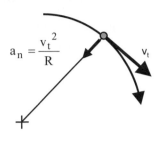

Now, since this is a real acceleration, it
must arise due to the application of a force.

Consider a stone
being whirled around
using a string. The normal
acceleration is v_t^2/r. This acceleration
must arise due to the application of an unbalanced force.
The force must also be in the same direction as the acceleration. Hence,
there is an unbalanced force directed inwards towards the centre of
rotation! We call this force a **centripetal force**.

The force is proportional to the mass
and the acceleration, thus:

$$F_C = \frac{mv_t^2}{R}$$

but $\quad v_t = R\omega$

thus $\boxed{F_C = m\omega^2 R}$

It is very convenient to include an **inertia force** in rotation problems so
that equilibrium conditions may may calculated. The inertia force F_I is
equal and opposite to that of the centripetal force. The inertial force is
often called the **centrifugal force**.

4.6.4 Planetary motion

The rotation of planets around the sun generates **centrifugal forces** which balance the inwards pulling gravitational force. The closer the planet is to the sun, the stronger the gravitational force (by the inverse square law) and the faster the **orbital velocity** must be to maintain its orbit.

Consider the Earth and the Sun:

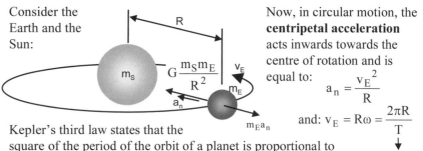

Now, in circular motion, the **centripetal acceleration** acts inwards towards the centre of rotation and is equal to:

$$a_n = \frac{v_E^2}{R}$$

and: $v_E = R\omega = \frac{2\pi R}{T}$

$\downarrow$
Period

Kepler's third law states that the square of the period of the orbit of a planet is proportional to the cube of planet's distance from the sun. Thus, since:

$$G\frac{m_S m_E}{R_E^2} = m_E \frac{4\pi^2 R_E}{T_E^2}$$

then: $\left[\dfrac{R_E^3}{T_E^2}\right] = G\dfrac{m_S}{4\pi^2}$

Since G is a constant which can be determined by experiment, the mass of the sun can be determined from observations of R and T of any planet.

Alternately, since all the terms on the right side of the equation above are constant, then the ratio of the mass of a planet with the mass of the sun can be calculated:

$$\left[\frac{R_m^3}{T_m^2}\right] = G\frac{m_p}{4\pi^2}$$

Now, by forming the ratio:

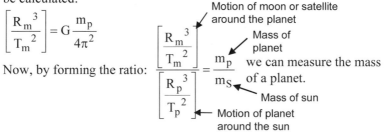

Motion of moon or satellite around the planet
Mass of planet
we can measure the mass of a planet.
Mass of sun
Motion of planet around the sun

For the motion of a satellite of mass m_1 around the Earth, the same principles apply and we have:

$$F = G\frac{m_1 m_E}{R^2} = \frac{m_1 v^2}{R}$$

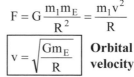

$$v = \sqrt{\frac{Gm_E}{R}}$$ **Orbital velocity**

4.6.5 Example

1. What is the maximum speed that a car may travel at without skidding on a level curve of radius 100 m for a coefficient of friction of 0.2 (raining) and 0.6 (dry)?

Solution:

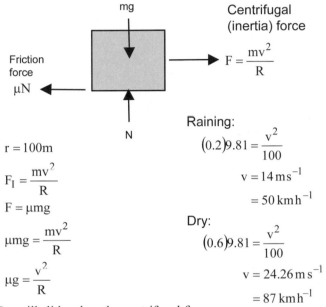

$$mg$$

Friction force
$$\mu N$$

$$N$$

Centrifugal (inertia) force

$$F = \frac{mv^2}{R}$$

$$r = 100\text{m}$$

$$F_I = \frac{mv^2}{R}$$

$$F = \mu mg$$

$$\mu mg = \frac{mv^2}{R}$$

$$\mu g = \frac{v^2}{R}$$

Raining:

$$(0.2)9.81 = \frac{v^2}{100}$$

$$v = 14\,\text{m s}^{-1}$$

$$= 50\,\text{km h}^{-1}$$

Dry:

$$(0.6)9.81 = \frac{v^2}{100}$$

$$v = 24.26\,\text{m s}^{-1}$$

$$= 87\,\text{km h}^{-1}$$

Car will slide when the centrifugal force becomes equal to the friction force (which is the centripetal force in this case).

4.7 Rotation

Summary

$$T = Fr$$ Torque

$$I = \sum mr^2$$ Moment of inertia about axis

$$M_I = I\alpha$$ Inertia moment

$$I_X = I_C + Md^2$$ Transfer formula

$$K = \sqrt{\frac{I}{M}}$$ Radius of gyration

4.7.1 Moment of inertia

Consider a wheel mounted on a shaft. The weight of the wheel is balanced by N. A horizontal force applied to the centre bearing of the assembly will simply cause the wheel to translate. However, if the force is applied to a point off-centre, such as at the rim, then the wheel will rotate about its axis.

The application of a force acting through a perpendicular distance in this manner is called a **moment** and sometimes referred to as a **torque**. The magnitude of the accelerating moment, or torque is:

$$T = Fr \longleftarrow$$ Perpendicular distance between the line of action of the force and the axis about which rotation occurs.

When the disk rotates under the action of an accelerating moment, it acquires **angular acceleration**. As in all our problems on dynamics, we introduce an **inertia force**, or in this case, an inertia moment, to act in the opposite sense to the accelerating moment. This allows rotational dynamics problems to be treated using equations of equilibrium.

What is the value of M_I? Since it acts in the opposite sense to M, then the magnitude is just Fr. However, we can express this **inertia moment** in terms of the **angular acceleration** just as we expressed the inertia force in translational problems using the product $F_I = ma$.

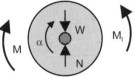

We divide the rotating mass into a large number of elements, each with a mass m. Each mass element has the same angular acceleration α, but is located at a particular radius from the axis of rotation r.

Now, the tangential acceleration of each mass element is: $a = r\alpha$

Thus, the inertia force acting on each mass element is: $F_I = ma = mr\alpha$

The first mass element is located a distance r_1 from the centre of rotation, then the inertia moment for this mass element is:

$$M_{I1} = F_I r_1$$
$$= mr_1{}^2\alpha$$

The total **inertia moment** M_I is the sum of all these:

$$M_I = \sum F_I r$$

The quantity $\sum mr^2$ is called the **moment of inertia** of the rotating body and has the units kg m².

$$= \left[\sum mr^2\right]\alpha$$

$$\boxed{M_I = I\alpha}$$

4.7.2 Moment of inertia of common shapes

The **moment of inertia** is closely related to the **2nd moment of an area**. The only difference is that here we are dealing with the masses rather than areas. The moment of inertia of a body about some axis can be determined by dividing the body into elemental masses and summing the resulting moments.

$$I = \sum mr^2$$

$$= \int r^2 dm$$

Formulas for moments of inertia of bodies of simple shapes are usually to be found in engineering data books.

Units: kg m²

The **moment of inertia of an area** has unit m⁴. The **moment of inertia of a body** is an entirely different thing and has units kg m².

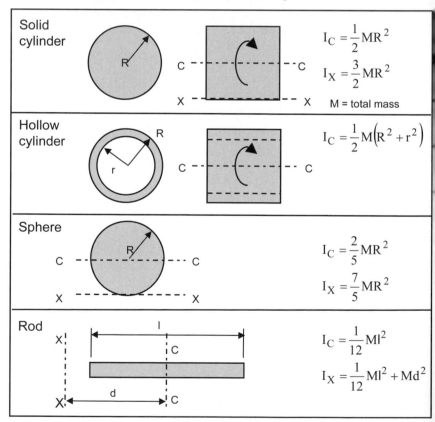

Solid cylinder	$I_C = \frac{1}{2}MR^2$
	$I_X = \frac{3}{2}MR^2$
	M = total mass

| Hollow cylinder | $I_C = \frac{1}{2}M\left(R^2 + r^2\right)$ |

| Sphere | $I_C = \frac{2}{5}MR^2$ |
| | $I_X = \frac{7}{5}MR^2$ |

| Rod | $I_C = \frac{1}{12}Ml^2$ |
| | $I_X = \frac{1}{12}Ml^2 + Md^2$ |

4.7.3 Composite bodies

The total moment of inertia of an irregularly shaped body can be often determined by dividing the body into simple shapes and using a **transfer formula**. For each simple shape, the moment of inertia about its own parallel **centroidal axis** can be found in reference tables. This moment of inertia is then transferred to the desired axis by:

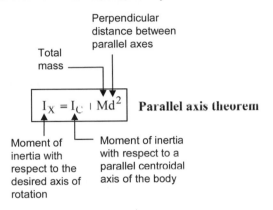

For example, in the preceding table, it can be seen that the moment of inertia about some distant axis XX is equal to the moment of inertia of the rod about its central axis plus the total mass times the distance to the axis squared.

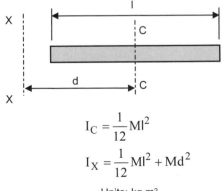

$$I_C = \frac{1}{12} M l^2$$

$$I_X = \frac{1}{12} M l^2 + M d^2$$

Units: kg m²

4.7.4 Radius of gyration

The **moment of inertia** of a body is a measure of its resistance to angular acceleration just as the mass of a body is a measure of its resistance to a translational acceleration.

Now, in our derivation of the formula for calculating the moment of inertia, we agreed that a body may be subdivided into a series of elemental masses each with its own radius from the centre of rotation and that the total moment of inertia is found from:

$$I = \sum mr^2$$

For example, for a solid disk, the moment of inertia around its central axis evaluates to $\frac{1}{2}MR^2$

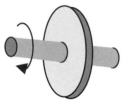

Now let us imagine another disk which has the same mass, but all of this mass is concentrated at a band of material at the rim of the disk. The radius of this disk is selected such that the moments of inertia of the two disks are the same and is given the symbol K The moment of inertia is (by inspection):

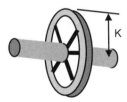

$$I_C = MK^2 \quad \text{All the mass M is at a distance K from the axis of rotation}$$

This radius K has a special meaning and is called the **radius of gyration**. It is the radius at which the total mass of a body may be evenly concentrated so as to obtain the same moment of inertia.

$$K = \sqrt{\frac{I}{M}}$$

If a large moment of inertia is required (such as in a flywheel), then the radius of gyration should be made as large as possible. This can be done without increasing the radial dimensions of the disk by concentrating the mass at the outer edge – i.e. by making the edge thicker.

4.7.5 Calculation of moment of inertia

Now, the centre of mass of an object is that point at which the total mass of the object can be said to be concentrated. We might think therefore that the moment of inertia of a body with respect to some axis can be obtained by simply multiplying the total mass by the distance from the centre of mass to that axis. Consider the following calculation for the moment of inertia of a beam about its end:

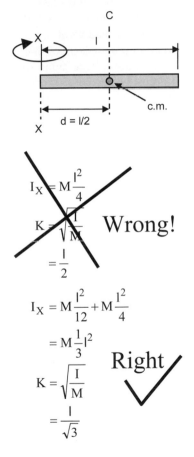

$$I_X = M\frac{l^2}{4}$$

$$K = \sqrt{\frac{I}{M}} \quad \text{Wrong!}$$

$$= \frac{l}{2}$$

Doing it this way ignores the effect of the distribution of the mass within the body about the axis in question. The r^2 factor in the definition of moment of inertia means that mass at a greater distance away from the axis makes a higher contribution to the overall value of I compared to mass closer to the axis.

$$I_X = M\frac{l^2}{12} + M\frac{l^2}{4}$$

$$= M\frac{1}{3}l^2$$

$$K = \sqrt{\frac{I}{M}} \quad \text{Right} \checkmark$$

$$= \frac{l}{\sqrt{3}}$$

Note, the **radius of gyration** is further away from the axis of rotation than the centre of mass showing that mass at a greater distance away from the axis contributes more to the **rotational inertia** of the body.

4.7.6 Dynamic equilibrium

In analysing rotational motion, it is convenient to use equilibrium equations. Thus, we consider an **inertia moment** which acts in the opposite sense to the **accelerating moment** - or **torque**.

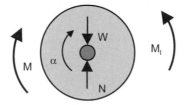

The **rotational inertia** of a body depends not only on its mass but on how this mass is distributed around the axis of rotation. The **radius of gyration** is a measure of this distribution. A large radius of gyration means that the mass of a rotating body has is distributed more towards the outer edges of the structure. A small radius of gyration means that the mass is concentrated more towards the centre of rotation.

Now, the inertia moment is equal to:

$$M_I = I\alpha$$

Notice the resemblance to the inertia force in translational problems. The moment of inertia I is the rotational equivalent of mass. The angular acceleration is the rotational equivalent of translational acceleration.

In problems involving rotation, we employ a **rotational dynamic equilibrium** by taking into consideration an inertia moment :

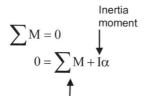

Inertia moment

$$\sum M = 0$$

$$0 = \sum M + I\alpha$$

Sum of all accelerating and decelerating moments (eg friction moments)

In these formulas, we must be careful to use signs correctly. We could settle on any particular system as long as we are consistent. For example, we could agree that clockwise moments are positive. Or, we could say that the direction of any accelerating moments or applied torques are positive and friction and inertia moments are negative.

4.7.7 Rolling motion

Plane motion, such as rolling without slipping, can
be treated as a combination of rotational and
translational motion. Consider a wheel being
accelerated by a pull P through its centre
of mass.

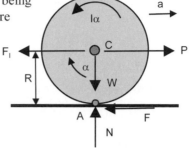

Translation

F is a friction force which is a reaction to the pull P acting on the centre of
mass. F_I is the inertia force given by: $F_I = Ma$

In this example, the action of force P causes the
disk to move to the right with an acceleration a.
An inertia force $F_I = ma$ acts towards the left.

$$\sum F_H = 0$$
$$0 = P - F_I - F$$
$$P = ma + \mu N$$

Rotation

An inertia moment $I\alpha$ balances the applied
torque arising from the friction force F
through perpendicular distance R.

$$\sum M_C = 0$$
$$0 = FR - M_I$$
$$FR = I\alpha$$

Note, the horizontal
forces P and F_i pass
through the centre of
mass and thus do not
contribute a moment
about the centre of mass.

Now, the disk has an angular acceleration α.
The tangential acceleration of a point on the
rim of the disk, such as point A, is: $a = R\alpha$

But, the acceleration of A with respect to the centre of mass C is exactly
the same in magnitude as the acceleration of C with respect to the point A.
That is, the horizontal translational acceleration of the centre of mass of the
disk. Hence, the translational inertia force F_I can be expressed in terms of
the angular acceleration: $F_I = Ma = MR\alpha$

4.7.8 Examples

1. Determine the moment of inertia about the central axis of a 8 spoked wagon wheel:

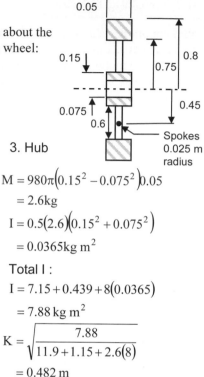

Solution: $\rho = 980$ kg m^{-3}

1. Rim

$M = 980\pi\left(0.8^2 - 0.75^2\right)0.05$

$= 11.9$kg

$I = 0.5(11.9)\left(0.8^2 + 0.75^2\right)$

$= 7.15$ kg m^2

2. Spokes

$M = 980\pi\left(0.025^2\right)0.6$

$= 1.15$kg

$I = \dfrac{1.15\left(0.6^2\right)}{2} + 1.15\left(0.45^2\right)$

$= 0.439$ kg m^2 (one spoke)

3. Hub

$M = 980\pi\left(0.15^2 - 0.075^2\right)0.05$

$= 2.6$kg

$I = 0.5(2.6)\left(0.15^2 + 0.075^2\right)$

$= 0.0365$kg m^2

Total I :

$I = 7.15 + 0.439 + 8(0.0365)$

$= 7.88$ kg m^2

$K = \sqrt{\dfrac{7.88}{11.9 + 1.15 + 2.6(8)}}$

$= 0.482$ m

2. A mass of 5 kg on the end of a rope wound around a drum results in a moment that is just sufficient to overcome friction in the bearings. The 300 kg drum then rotates at a uniform angular velocity. Determine the coefficient of friction if the shaft diameter is 0.15 m.

Solution:

Free body diagram

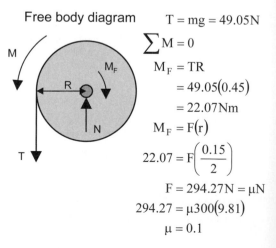

$T = mg = 49.05$N

$\sum M = 0$

$M_F = TR$

$= 49.05(0.45)$

$= 22.07$Nm

$M_F = F(r)$

$22.07 = F\left(\dfrac{0.15}{2}\right)$

$F = 294.27$N $= \mu$N

$294.27 = \mu 300(9.81)$

$\mu = 0.1$

3. To determine the moment of inertia of the drum shown in (2), a mass of 25 kg is attached to the rope around the rim and allowed to fall. The friction moment M_F is 22 N m. If the mass is observed to drop a distance of 4 m in 4 seconds, calculate the moment of inertia about the central axis.

Solution:

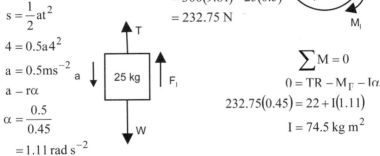

$$s = \frac{1}{2}at^2$$

$$4 = 0.5a4^2$$

$$a = 0.5 \, \text{ms}^{-2}$$

$$a - r\alpha$$

$$\alpha = \frac{0.5}{0.45}$$

$$= 1.11 \, \text{rad s}^{-2}$$

$$T = mg - F_I$$
$$= 300(9.81) - 25(0.5)$$
$$= 232.75 \, \text{N}$$

$$\sum M = 0$$
$$0 = TR - M_F - I\alpha$$
$$232.75(0.45) = 22 + I(1.11)$$
$$I = 74.5 \, \text{kg m}^2$$

4. A motor mechanic presses the end of a tool against a 200 mm diameter, 20 mm thick grindstone which is rotating at 3000 rpm. The force applied to the grindstone at its outer edge is 100 N. The power to the grindstone is suddenly shut off. Calculate how long it takes for the grindstone to come to a rest (ignoring friction at the grindstone centre bearing).

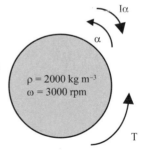

Solution:

$$I = \frac{1}{2}MR^2$$

$$= \frac{1}{2}(\pi 0.1^2)(0.02)(2000)0.1^2$$

$$= 0.0063 \, \text{kg m}^2$$

$$T = I\alpha$$
$$-100(0.1) = (0.0063)\alpha$$
$$\alpha = -1587 \, \text{rad s}^{-2}$$
$$0 = \left(3000\frac{2\pi}{60}\right)^2 + -1587t$$
$$t = 62 \, \text{s}$$

4.8 Work and energy

Summary

$$W = Fs \qquad \text{Work}$$

$$F = -ks \qquad \text{Spring restoring force}$$

$$W = -\frac{1}{2}ks^2 \qquad \text{Work done on/by a spring}$$

$$PE = mgh \qquad \text{Gravitational potential energy}$$

$$PE = \frac{1}{2}ks^2 \qquad \text{Strain potential energy}$$

$$KE = \frac{1}{2}mv^2 \qquad \text{Kinetic energy}$$

$$KE = \frac{1}{2}I\omega^2 \qquad \text{Rotational kinetic energy}$$

$$W = T\theta \qquad \text{Rotational work}$$

$$P = Fv \qquad \text{Power}$$

$$P = T\omega \qquad \text{Rotational power}$$

4.8.1 Work and energy

- Energy exists in many forms
- There is a flow, or transfer, of energy when a change of form takes place
- **Heat** and **work** are words which refer to the amount of <u>energy in transit</u> from one place to another.
- Heat and Work cannot be stored.

We generally speak of two types of mechanical energy:

Potential energy	**Kinetic energy**
The energy associated with the position of a body in the **gravitational field** or that stored within a body placed under stress.	The energy associated with the velocity of a body.

Kinetic and potential energy may be converted from one type to another. The transfer occurs through the mechanism of **work**.

The application of an external force acting through a distance signifies a transfer of energy which we call **work**. Work is done <u>by</u> forces <u>on</u> a body.

The magnitude of the work done W on a body by force F is the product of the force F and the distance s through which it acts.

$$W = Fs$$

The units of work are Newtons metres. Since work is energy in transit from one form to another, the units of work are the same as energy: Nm = **Joules**

Work may be positive or negative. When the direction of the force is in the same direction as the displacement of the body, the work is positive.

Note, we use the symbol W for both Work and Weight. The context should make it clear which is meant.

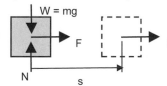

If the direction of force is same as that of direction of the displacement "s", then the work done by F is positive.

The forces W and N do not do any work as the block is moved horizontally along the surface by the force P and there is no displacement in the vertical direction. Work is done <u>BY</u> forces <u>ON</u> bodies. When the force acts in the same direction as the displacement of the body, the work done is positive. When the force acts in the opposite direction as the displacement of the body, the work done is negative.

4.8.2 Variable force

A good example of a variable force is that offered by a compressed or stretched spring.

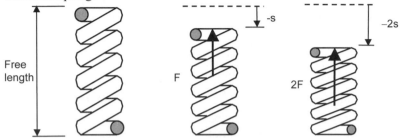

When a spring is compressed, it provides a resisting or **restoring force** whose magnitude is proportional to the amount of compression. The constant of proportionality is called the **spring constant** or **spring stiffness**.

The constant of proportionality if called the spring "stiffness" and has the units N/m

$$F = -ks$$

The minus sign means that the restoring force F is opposite in direction to the displacement s.

As the spring is compressed, it offers an ever increasing restoring force. If, in compressing the spring from a displacement = 0 to −s, we break the compression up into a series of small steps over which we consider the the **restoring force** to be more or less constant, then adding the work done for each of these small steps together gives us the work done <u>by</u> the spring.

$$W = \sum F \Delta s$$

If the steps are made very very small, then mathematically, we can state:

$$W = \int_0^{-s} F ds$$

$$= \int_0^{-s} -ks\,ds$$

$$W = -\frac{1}{2}ks^2$$

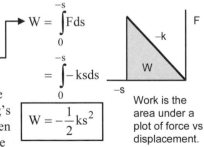

Work is the area under a plot of force vs displacement.

Is the work positive or negative? We speak about a spring doing work <u>on</u> the body which is deforming it. The spring's restoring force does negative work when being compressed or stretched since the direction of the restoring force is always opposite to that of the displacement. When the spring is released, it does positive work on the body to which it is attached since now, the force provided by the spring is in the same direction as the displacement of the body.

4.8.3 Work done by a spring

Up: (+)   Down: (−)   Left: (−)   Right: (+)

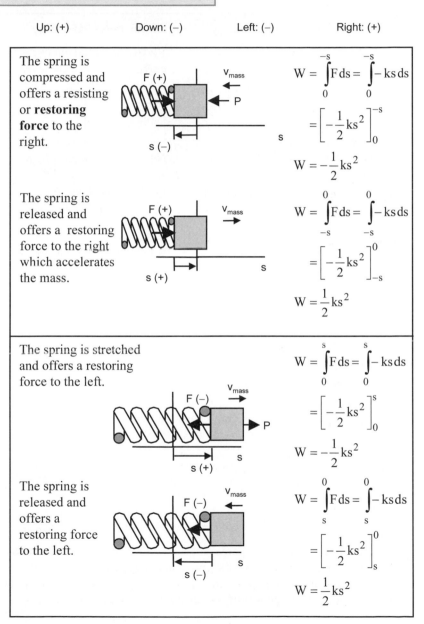

The spring is compressed and offers a resisting or **restoring force** to the right.

$$W = \int_0^{-s} F\,ds = \int_0^{-s} -ks\,ds$$

$$= \left[-\frac{1}{2}ks^2 \right]_0^{-s}$$

$$W = -\frac{1}{2}ks^2$$

The spring is released and offers a restoring force to the right which accelerates the mass.

$$W = \int_{-s}^{0} F\,ds = \int_{-s}^{0} -ks\,ds$$

$$= \left[-\frac{1}{2}ks^2 \right]_{-s}^{0}$$

$$W = \frac{1}{2}ks^2$$

The spring is stretched and offers a restoring force to the left.

$$W = \int_0^{s} F\,ds = \int_0^{s} -ks\,ds$$

$$= \left[-\frac{1}{2}ks^2 \right]_0^{s}$$

$$W = -\frac{1}{2}ks^2$$

The spring is released and offers a restoring force to the left.

$$W = \int_s^{0} F\,ds = \int_s^{0} -ks\,ds$$

$$= \left[-\frac{1}{2}ks^2 \right]_s^{0}$$

$$W = \frac{1}{2}ks^2$$

4.8.4 Energy

Gravitational potential energy

The energy associated with the position of a body.

Gravitational potential energy of a body may be increased by raising it a distance h. When a body moves to a lower level, its gravitational potential energy is reduced.

The raising or lowering of a body is a force mg being applied through a distance h. Thus, the work or change in potential energy is:

$$PE = mgh$$

Elastic potential energy

Elastic potential energy is the energy stored in a stretched or compressed spring.

$$PE = \frac{1}{2}ks^2$$

Note: The work done <u>by</u> a spring when released is positive and given by:

$$W = \frac{1}{2}ks^2$$

Work done <u>by</u> spring when being compressed or stretched is negative:

$$W = -\frac{1}{2}ks^2$$

Kinetic energy

The energy associated with the velocity of a body.

Consider an unbalanced force P applied to a mass initially at rest on a frictionless surface. The mass is given an acceleration a and moves a distance s and acquires a velocity v.

$$v^2 = 2as$$
$$U = Fs$$
$$= mas$$

$$KE = \frac{1}{2}mv^2$$

The correspondence between work and energy is a powerful tool in solving mechanics problems. The procedure for applying the energy approach is:

| Initial kinetic energy of a body | + Positive work | − Negative work | | = | Final kinetic energy |

This equation is another way of saying that energy is neither created or destroyed but only changed into another form by the mechanisms of either heat or work. This is the **law of conservation of energy**.

4.8.5 Rotational kinetic energy

The moment of inertia of a body is a measure of its resistance to angular acceleration just as the mass of a body is a measure of its resistance to a translational acceleration.

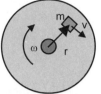

We divide the rotating mass into a large number of elements, each with a mass m. Each mass element has the same angular velocity ω, but is located at a particular radius from the axis of rotation r.

Now, the kinetic energy of each mass element is: $KE = \frac{1}{2}mv^2$

But, $v = r\omega$, thus the kinetic energy for a single mass element becomes: $KE = \frac{1}{2}mr^2\omega^2$

The total kinetic energy is the sum of all these: $KE = \sum \frac{1}{2}mr^2\omega^2$

But, the quantity $\sum mr^2$ is the **moment of inertia** of the rotating body.

$$= \frac{1}{2}\left[\sum mr^2\right]\omega^2$$

Thus, the **rotational kinetic energy** is: $\boxed{KE = \frac{1}{2}I\omega^2}$

The work done on a rotating body can be considered in terms of the moment or **torque** "T" applied moving through an angular displacement θ

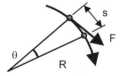

$W = Fs$

$= FR\theta$ since $s = R\theta$

$\boxed{W = T\theta}$

	Translational	Rotational
Distance	s	θ
Velocity	v	ω
Acceleration	a	α
Mass	m	I
Work	W = Fd	W = Tθ
Kinetic energy	$\frac{1}{2}mv^2$	$\frac{1}{2}I\omega^2$

4.8.6 Power

Consider a mass m that has to be raised to a height h.

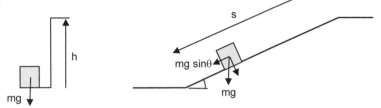

The amount of work to be done is W = mgh. Now, to raise this mass, we can apply a force equal to mg through a height h, or we can use an inclined slope and apply a smaller force F = mg sinθ over a longer distance s. The product Fs comes out to be equal to mgh since the same amount of work has to be done no matter how the mass is raised (ignoring any friction on the slope). The difference between the two methods is that the second method takes a longer time and thus requires less power.

Power = time rate of doing work and is measured in Joules per second, or **Watts**.

$$P = \frac{dW}{dt}$$ Power = work over time

$$= \frac{dFs}{dt}$$

$$= F\frac{ds}{dt}$$ if F is a constant

$$P = Fv$$ Power = force times velocity

A very powerful machine is capable of doing a lot of work in a short time. The exact same amount of work may be able to be done by a less powerful machine but the time taken to do this work will be longer.

In rotational motion, power is still the time rate of change of doing work, thus:

$$P = \frac{dW}{dt}$$

$$= \frac{dT\theta}{dt}$$ where T is torque

$$= T\frac{d\theta}{dt}$$

$$P = T\omega$$

A small engine can deliver large amounts of power if it spins quickly enough. For example, a small engine which develops 50 Nm of torque at 10,000 rpm (1047 rad s⁻¹) is producing 52 kW of power. The same power may be had with a larger engine which develops more torque at a lower rpm. In the above example, 50 kW of power occurs at at 2400 rpm if the torque developed is 200 Nm. The engine in a cargo ship may develop a huge amount of torque but only be turning a few hundred rpm which, when multiplied together produces an enormous power output.

4.8.7 Relativity

Einstein found that classical concepts of length, time and mass required modification to satisfy the requirement for the general applicability of physical laws and the observed constant speed of light. In the **theory of special relativity**, he proposed that:

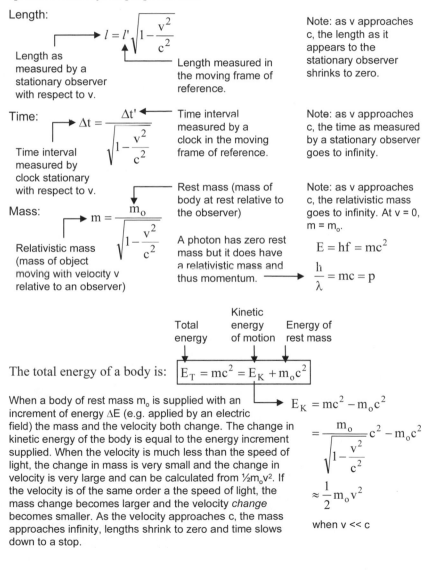

Length:

$$l = l'\sqrt{1 - \frac{v^2}{c^2}}$$

Length as measured by a stationary observer with respect to v.

Length measured in the moving frame of reference.

Note: as v approaches c, the length as it appears to the stationary observer shrinks to zero.

Time:

$$\Delta t = \frac{\Delta t'}{\sqrt{1 - \frac{v^2}{c^2}}}$$

Time interval measured by clock stationary with respect to v.

Time interval measured by a clock in the moving frame of reference.

Note: as v approaches c, the time as measured by a stationary observer goes to infinity.

Mass:

$$m = \frac{m_o}{\sqrt{1 - \frac{v^2}{c^2}}}$$

Relativistic mass (mass of object moving with velocity v relative to an observer)

Rest mass (mass of body at rest relative to the observer)

A photon has zero rest mass but it does have a relativistic mass and thus momentum. ⟶

Note: as v approaches c, the relativistic mass goes to infinity. At v = 0, m = m_o.

$$E = hf = mc^2$$

$$\frac{h}{\lambda} = mc = p$$

The total energy of a body is:

Total energy | Kinetic energy of motion | Energy of rest mass

$$E_T = mc^2 = E_K + m_o c^2$$

When a body of rest mass m_o is supplied with an increment of energy ΔE (e.g. applied by an electric field) the mass and the velocity both change. The change in kinetic energy of the body is equal to the energy increment supplied. When the velocity is much less than the speed of light, the change in mass is very small and the change in velocity is very large and can be calculated from $\frac{1}{2}m_o v^2$. If the velocity is of the same order a the speed of light, the mass change becomes larger and the velocity *change* becomes smaller. As the velocity approaches c, the mass approaches infinity, lengths shrink to zero and time slows down to a stop.

$$E_K = mc^2 - m_o c^2$$

$$= \frac{m_o}{\sqrt{1 - \frac{v^2}{c^2}}} c^2 - m_o c^2$$

$$\approx \frac{1}{2} m_o v^2$$

when v << c

4.8.8 Examples

1. The driver of a car brakes hard to avoid collision with a pedestrian. Skid marks on the road indicate that the car was brought to rest in 15 m. There is disagreement between the driver and other witnesses as to how fast the car was travelling initially. Tests with a similar car show that the coefficient of friction between the tyres and the road surface is 0.6. The mass of the car is 1600 kg. Determine the initial velocity of the car.

Solution:

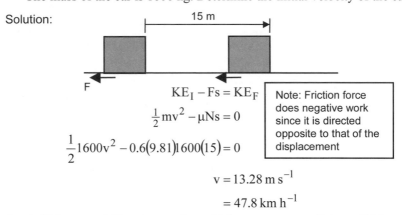

$$KE_I - Fs = KE_F$$

$$\tfrac{1}{2}mv^2 - \mu Ns = 0$$

$$\tfrac{1}{2}1600v^2 - 0.6(9.81)1600(15) = 0$$

Note: Friction force does negative work since it is directed opposite to that of the displacement

$$v = 13.28 \text{ m s}^{-1}$$

$$= 47.8 \text{ km h}^{-1}$$

2. A 10 kg mass falls onto a spring which compresses a distance 0.15 m. The spring stiffness is 7200 N m^{-1}. What height above the free length of the spring was the mass positioned initially?

Solution:

Note: The mass starts and finishes from rest hence its initial and final kinetic energy is zero. The force offered <u>by</u> the spring <u>on</u> the mass does negative work since the spring force is in the opposite direction to the displacement of the mass.

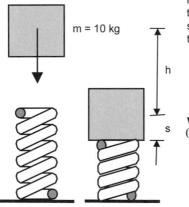

m = 10 kg

Initial KE   negative
positive work
work   Final KE

$$0 + mg(h+s) - \tfrac{1}{2}ks^2 = 0$$

$$10(9.81)(h + 0.15) = \tfrac{1}{2}7200(0.15^2)$$

$$h = 0.68 \text{ m}$$

3. A yo-yo is allowed to fall under its own weight. The string around the rim causes the yo-yo to rotate while falling. The other end of the string is held fixed. What is the translational acceleration of the yo-yo? Employ the energy method for the solution.

Solution:

The yo-yo has the shape of a disk.

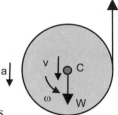

Let us first consider the rotational motion. The disk is acted upon by a moment P×R which tends to increase the angular velocity of the disc. The product of the moment and the angular displacement thus represents positive work done on the disk.

Note: In energy problems, we do NOT show internal inertia forces in free body diagrams. These were tools we used for the force method of analysis.

$$W = [PR]\theta$$
$$s = R\theta$$
$$W = Ps$$

Let the yo-yo fall a distance s = h. The force W = mg is acting in the same direction as h, thus, the positive work done is mgh. The force T acts in the opposite direction through a distance h and negative work is done. The initial kinetic energy of the disk is zero, and the final kinetic energy is the sum of the translational and rotational KE's.

$$mgh - Ph + Ph = \frac{1}{2}mv^2 + \frac{1}{2}I\omega^2$$

$$mgh = \frac{1}{2}\left[mv^2 + \frac{1}{2}mR^2\frac{v^2}{R^2}\right]$$

$$= \frac{1}{2}m\left[\frac{3}{2}v^2\right] = \frac{3}{4}mv^2$$

$$v^2 = \frac{4}{3}gh = 2ah$$

$$a = \frac{2}{3}g \quad \text{Answer}$$

4.9 Impulse and momentum

Summary

$$p = mv \qquad \text{Momentum}$$

$$\frac{\Delta p}{\Delta t} = ma \qquad \text{Rate of change of momentum}$$

$$I = Ft \qquad \text{Impulse}$$

$$\text{Angular momentum} = I\omega \qquad \text{Angular momentum}$$
$$\text{Angular impulse} \quad = Tt \qquad \text{Angular impulse}$$

$$m_1 v_1 + m_2 v_2 = m_1 v_1' + m_2 v_2' \qquad \text{Conservation of momentum}$$

4.9.1 Impulse and momentum

The concepts of **impulse** and **momentum** are yet more tools for solving problems in mechanics.

Momentum is a property of a body that determines the length of time it takes to bring it to rest. It is equal to the product of the mass and the velocity of the body and is given the symbol p.

$$p = mv$$

If a body is in motion, then of course the length of time taken to bring it to rest depends on the magnitude of the force we use to slow it down.

Momentum is a vector quantity (since velocity is a vector quantity). The units of momentum are kg m s^{-1}

Let us imagine a mass m moving with velocity v_o in a horizontal direction to the right. A retarding force is applied which eventually brings the mass to rest.

Initially, the momentum of the body is mv_o. After a time interval Δt, the velocity has decreased to say v_1. There has been a change in momentum during the time Δt because the velocity has changed $\Delta v = v_o - v_1$. The change in momentum is: $\Delta p = m\Delta v$

The rate of change of momentum with time is: $\dfrac{\Delta p}{\Delta t} = \dfrac{m\Delta v}{\Delta t}$

$$= ma$$

But, the product ma is the force acting on the body. Thus we can say that Force is actually the **rate of change of momentum**.

1. Momentum is something that the body possesses by virtue of its motion.	2. We apply forces for a certain time to the body to change its momentum.	3. A larger force changes the momentum more compared to a smaller force for the same time of application.

Momentum is a measure of the time taken to bring a body to rest with a given force. Thus, to bring a body to rest, we apply a force F for a time t. The product of force times time is called **impulse**.

Momentum is something that belongs to a body - it is a measure of its **motion**.

$$I = Ft$$

Impulse is something that we apply to a body to change its momentum.

4.9.2 Impulse-momentum equation

In solving mechanics problems by the impulse-momentum method, we consider the initial momentum of a body, the positive and negative impulses upon it, and thus obtain the final momentum.

Impulse-momentum equation

| Initial momentum | + | Positive impulses | − | Negative impulses | | = | Final momentum |

This equation must be applied with careful attention to the directions of the forces and velocities involved. We must decide on a positive direction for velocities. Only components of force acting in the same direction as the velocities contribute to the momentum and impulse given a body.

The momentum associated with the rotational motion of a body is called **angular momentum**. Angular momentum is a property of a rotating body which determines the time it takes to bring the body to rest.

Angular Impulse - momentum equation

| Initial angular momentum | + | Positive angular impulses | − | Negative angular impulses | | = | Final angular momentum |

In rotational motion, mass is replaced by **moment of inertia** and **force** by **torque**. Thus:

$$\text{Angular momentum} = I\omega$$
$$\text{Angular impulse} = Tt$$

Momentum is not **energy**. Momentum is force × time, not force × distance. Momentum is a measure of the <u>time</u> required to bring a body to rest, not a measure of the <u>distance</u> through which a force acts to bring a body to rest.

Imagine that a force F is applied to a large ship moving at a low velocity. The product of the mass times the velocity of the ship is a measure of the time through which the force F must be applied to bring it to rest. This results in the ship travelling a certain distance s during this time.

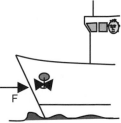

If the initial velocity of the ship was doubled, the ship has now got four times the initial **kinetic energy** and if the same force is used, then four times the distance would be required since W = Fs. But, the ship has only *twice* the **momentum**, and thus the time through which the force is to be applied is only doubled.

4.9.3 Conservation of momentum

Collisions which result in no dissipation of energy are called **elastic**.
Collisions involving dissipation of energy are called **inelastic**.

(a) two bodies move towards each other.

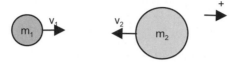

(b) they collide elastically imparting a force F to each other during a
time t

(c) and then move away from each other

Applying the **impulse-momentum equation**, we have:

$$m_1 v_1 + m_2(-v_2) + Ft - Ft = m_1(-v_1') + m_2 v_2'$$

$$\boxed{m_1 v_1 + m_2 v_2 = m_1 v_1' + m_2 v_2'}$$

Where we have written the velocity as vector
quantities to remind us to consider the
directions of the velocities involved.

Note that the two impulses have
cancelled out since the force acting
between the bodies is equal and opposite
in direction (by Newton's third law).

This **law of conservation
of momentum** applies to
inelastic collisions as
well as collisions which
are not head-on but
oblique or glancing - as
long as the velocity
vectors are treated
correctly.

A collision is a process in which momentum may be transferred from one
body to another but the total momentum of the system is unchanged.

4.9.4 Examples

1. A 1600 kg car moving with velocity 24 km h^{-1} strikes the rear end of a truck which was moving in the same direction (to the right) but at 10 km h^{-1}. The two vehicles become locked together during the collision. Determine the velocity of the vehicles after impact and change in KE.

Solution:

$$v_1 = \frac{24}{3.6}$$

$$= 6.67 \text{ m s}^{-1}$$

$$v_2 = \frac{10}{3.6}$$

$$= 2.78 \text{ m s}^{-1}$$

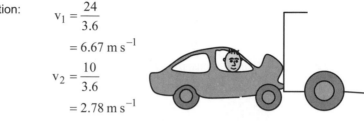

$$m_1 v_1 + m_2 v_2 = (m_1 + m_2)v$$
$$1600(6.67) + 8000(2.78) = 9600v$$

$$v = 3.43 \text{ m s}^{-1} = 12.3 \text{ km h}^{-1}$$

$$KE_I = \tfrac{1}{2}1600(6.67^2) + \tfrac{1}{2}8000(2.78^2) = 66.5 \text{ kJ}$$

$$KE_F = \tfrac{1}{2}(1600 + 8000)3.43^2 = 56.5 \text{ kJ}$$

$$\Delta KE = 66.5 - 56.5 = 10 \text{ kJ}$$ Energy lost to heat, deformation and sound during inelastic collision

2. If the collision was completely elastic, calculate the velocity of the car and the truck after impact. (Hint: No energy lost in elastic collisions)

Solution:

$$KE_I = \tfrac{1}{2}1600(6.67^2) + \tfrac{1}{2}8000(2.78^2)$$

$$= 66.5 \text{kJ}$$

$$= KE_F$$

$$66500 = \tfrac{1}{2}1600(v'_1{}^2) + \tfrac{1}{2}8000(v'_2{}^2)$$

$$= 800v'_1{}^2 + 4000v'_2{}^2 \quad ..... (1)$$

$$m_1 v_1 + m_2 v_2 = m_1 v'_1 + m_2 v'_2$$
$$1600(6.67) + 8000(2.78) = 1600v'_1 + 8000v'_2$$

$$1600v'_1 + 8000v'_2 = 32912 \quad ..... (2)$$ Now we have two equations (1) and (2), two unknowns v'_1

$$v'_1 = 0.187 \text{ ms}^{-1}$$ and v'_2, and these can be

$$v'_2 = 4.08 \text{ ms}^{-1}$$ solved by substitution.

3. A yo-yo is allowed to fall under its own weight. The string around the
   rim causes the yo-yo to rotate while falling. The other end of the string
   is held fixed. What is the translational acceleration of the yo-yo?
   Employ the impulse-momentum method for the solution.

Solution:

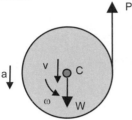

In this problem, equations for both linear and
angular momentum may be used to determine
the linear acceleration of the disk. First, let
the disk fall for a time t. Then, write out the
equations for both linear and angular
momentum separately and include the
appropriate positive and negative impulses.

For linear momentum, the
linear impulses are W and P
acting for a time t. For
angular momentum, the
angular impulse is the
moment P acting through R
around C for time t.

$$\text{Linear} \quad Mv_1 + Wt - Pt = Mv_2$$
$$(Mg - P)t = Mv_2$$
$$\text{Angular} \quad I\omega_1 + PRt = I\omega_2$$
$$PRt = I\omega_2$$

$$\text{but} \quad I = \frac{1}{2}MR^2$$

$$\text{and} \quad \omega = \frac{v_2}{R}$$

$$\text{thus} \quad PRt = \frac{1}{2}MR^2\frac{v_2}{R}$$

$$Pt = \frac{1}{2}Mv_2$$

$$\text{Substituting:} \quad Mgt - \frac{1}{2}Mv_2 = Mv_2$$

$$gt = \frac{3}{2}v_2$$

$$\frac{v_2}{t} = \frac{2}{3}g$$

$$\text{but} \quad v_2 = at$$

$$\text{thus} \quad a = \frac{2}{3}g$$

Part 5

Properties
of
Matter

5.1 Fluids

Summary

$p = p_g + p_{atm}$ Absolute pressure

$\Delta p = \rho g \Delta h$ Hydrostatic pressure

$F_2 = \dfrac{A_2}{A_1} F_1$ Pascal's principle

$\gamma = \dfrac{F_R}{d}$ Coefficient of surface tension

$\gamma = \dfrac{F}{2l} = \dfrac{W}{\Lambda A}$ Coefficient of surface tension

$\gamma = \dfrac{\Delta p R}{4}$ Surface tension - Bubble

$\gamma = \dfrac{\Delta p R}{2}$ Surface tension - Drop

$\gamma = \dfrac{\rho g h R}{2 \cos \theta}$ Contact angle

$A_1 v_1 = A_2 v_2$ Equation of continuity

$p_1 + \rho g h_1 + \dfrac{1}{2} \rho v_1^2 = p_2 + \rho g h_2 + \dfrac{1}{2} \rho v_2^2$ Bernoulli's equation

$\eta = \dfrac{F/A}{dv/dy}$ Viscosity

$F_v = 6 \pi \eta v R$ Stokes' law

$v_t = \dfrac{2 R^2 g (\rho_s - \rho_l)}{9 \eta}$ Terminal velocity

$Q = \dfrac{\pi \Delta p R^4}{8 \eta l}$ Flow rate through tube

5.1.1 Fluids

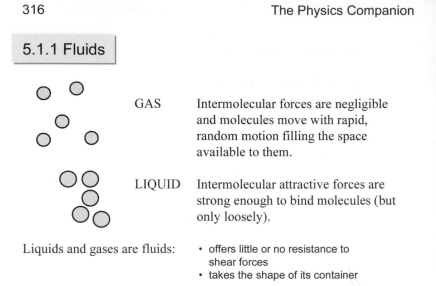

GAS   Intermolecular forces are negligible and molecules move with rapid, random motion filling the space available to them.

LIQUID   Intermolecular attractive forces are strong enough to bind molecules (but only loosely).

Liquids and gases are fluids:
- offers little or no resistance to shear forces
- takes the shape of its container

When dealing with fluids, some of the more important mechanical properties to be considered are the density, relative density, and the pressure.

Density

Mass in kg

$$\rho = \frac{m}{V}$$ (Units: kg m^{-3})

Volume m^3

Material	Density kg m^{-3}
Air	1.2
Water	1000
Lead	113000

Relative Density (or **specific gravity** - s.g.)

$$r.d. = \frac{\text{density of substance}}{\text{density of water}}$$ (No units)

Pressure

$$\text{Pressure} = \frac{\text{Force}}{\text{Area}}$$ (Units: N m^{-2} = Pa)

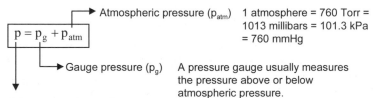

Atmospheric pressure (p_{atm})   1 atmosphere = 760 Torr = 1013 millibars = 101.3 kPa = 760 mmHg

$$p = p_g + p_{atm}$$

Gauge pressure (p_g)   A pressure gauge usually measures the pressure above or below atmospheric pressure.

Absolute pressure is usually used in most physics formulae

5.1.2 Hydrostatic pressure

There are two parts to **hydrostatic pressure**:

(1) The pressure due to the weight of the fluid itself.

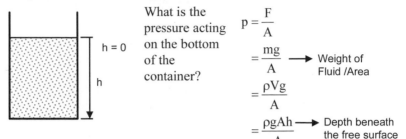

What is the pressure acting on the bottom of the container?

$$p = \frac{F}{A}$$

$$= \frac{mg}{A} \longrightarrow \text{Weight of Fluid /Area}$$

$$= \frac{\rho V g}{A}$$

$$= \frac{\rho g A h}{A} \longrightarrow \text{Depth beneath the free surface}$$

But, this is a **gauge pressure** (i.e. pressure above (or below) atmospheric pressure), thus the total **absolute pressure** is:

$$\boxed{p = \rho g h}$$

$$\boxed{p = p_o + \rho g h}$$

the pressure at the surface of the fluid (i.e. where h = 0)

Pressure only depends on density and depth beneath the surface of the fluid and not on the volume of fluid.

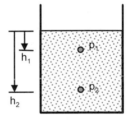

The pressure *difference* between two points at different depths is thus: $p_2 - p_1 = \rho g (h_2 - h_1)$

$$\boxed{\Delta p = \rho g \Delta h}$$

(2) The pressure which is applied by some external means

- Pressure applied to an enclosed fluid is transmitted equally in all directions throughout the fluid.

- The direction of pressure is always perpendicular to the area being considered but at any particular point in the fluid, is equal in all directions. **(Pascal's principle)**

Pascal's principle can be used to advantage in hydraulic systems using **pistons**. A small force applied to a small diameter piston of area A_1 can be connected, by fluid pressure, into a larger force at a larger diameter piston with area A_2. The force applied can be calculated from the ratio of the areas in the two pistons:

$$\boxed{F_2 = \frac{A_2}{A_1} F_1}$$

5.1.3 Archimedes' principle

1. Consider a glass of milk.

2. Consider a volume V of liquid within the milk.

3. Liquid V is subjected to an upwards force which equals the weight W of the liquid in V.

   The force must come from the surrounding fluid since there is nothing else in contact with the volume V. The force must equal the weight of the volume V since otherwise the volume V would be moving up or down.

4. If the liquid V were now replaced with a solid of exactly the same shape, then this solid would be subjected to exactly the same force as was the volume V of liquid. This force is equal to the weight of the liquid displaced.

Archimedes' principle

When a body is completely or partially immersed in a fluid, the fluid exerts an upwards **buoyancy force** equal to the weight of the fluid displaced.

Q. Will a body immersed in a fluid sink or float?

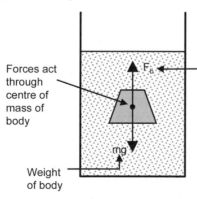

Buoyancy force or **"upthrust"** = weight of volume of liquid displaced

Forces act through centre of mass of body

Weight of body

A. Body will sink if $mg > F_b$
Body will float if $mg < F_b$

Q. What happens if a body is only partially submerged?

A. The body sinks until its weight is equal to the weight of the volume of water displaced. The greater the density of the fluid (e.g. salt water), the higher the body sits on the surface since less volume has to be displaced to match weight of body.

5.1.4 Surface tension

Forces between atoms or molecules take the form of a repulsion that is very strong at short distances and an attraction which diminishes in strength with larger distances. Atoms and molecules take up an equilibrium position where the **repulsive** and **attractive forces** are balanced.

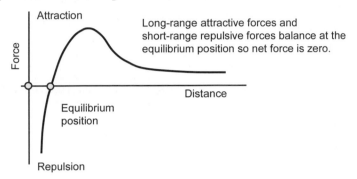

Consider two molecules in a liquid, one on the surface, and another in the interior. **Long-range** attractive forces F_A have a resultant R_A zero on A, downwards on B. **Short-range** repulsive forces F_R have a resultant R_R zero on A, upwards on B.

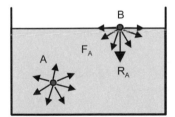
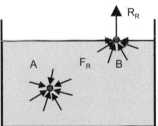

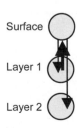

But, attractive forces are *long-range* forces repulsive forces are *short-range*, thus, a molecule B at the surface feels an attraction from the molecule at layers 1 and 2 since attractive forces are long range. $R_A = F_{A1} + F_{A2}$ However, the molecule at the surface only feels the repulsion from the molecule directly beneath it ($R_R = F_{R1}$) since repulsive forces are short range. Thus, to counterbalance all the "extra" attractive forces from the deeper molecules, the surface molecule has to move downwards and closer to layer 1 since the repulsive force increases with decreasing distance.

5.1.5 Contact angle

For any surface molecule, the **short-range repulsive forces** come from neighbours to the side and below but not from the top.

Resultant **long-range attractive force** must act in opposite direction so as to balance short-range forces and so are also perpendicular to the surface (on average)

or: the surface must be perpendicular to the resultant long-range attraction on the surface molecules.

Gravity acts vertically downwards.

All other forces involved will push the surface around until these conditions are met.

Resultant short-range forces are perpendicular to the surface by symmetry (on average)

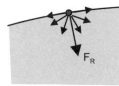

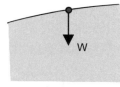

In a large pond, gravitational effects are dominant and pulls the fluid surface into a horizontal plane. Gravity acts on all molecules surface and interior. Gravity is thus a volume effect that depends on dimensions cubed. In a small droplet, surface tension effects are dominant and surface is pulled into a shape that minimises the surface area of the drop (i.e. a sphere). Surface tension only acts on surface molecules and thus is a surface effect and depends on linear dimension only.

These issues determine the **wetting angle** or **contact angle** of the liquid..

Case 1: **Wet contact**

Liquid molecules attracted more by the solid than liquid.

Case 2: **Waterproof contact**

Liquid molecules attracted more by the liquid than solid molecules.

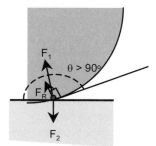

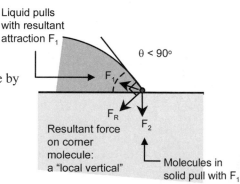

Liquid pulls with resultant attraction F_1

$\theta < 90°$

Resultant force on corner molecule: a "local vertical"

Molecules in solid pull with F_1

In both cases, the liquid is pulled into shape by surface forces until the surface of the liquid is perpendicular to the resultant force F_r.

5.1.6 Coefficient of surface tension

The coefficient of surface tension γ is defined as: $\boxed{\gamma = \dfrac{F_R}{d}}$ units of γ are N m^{-1}

Examples:

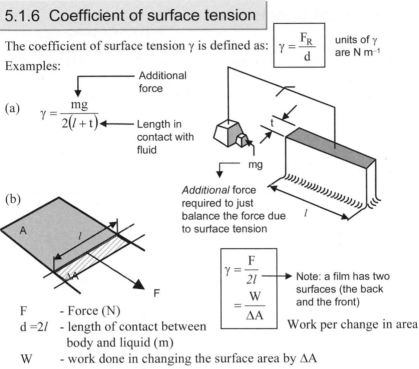

Additional force

(a) $\gamma = \dfrac{mg}{2(l+t)}$ ← Length in contact with fluid

mg

Additional force required to just balance the force due to surface tension

(b)

A

l

F

$\boxed{\gamma = \dfrac{F}{2l} = \dfrac{W}{\Delta A}}$ → Note: a film has two surfaces (the back and the front)

Work per change in area

F   - Force (N)
$d = 2l$ - length of contact between body and liquid (m)
W   - work done in changing the surface area by ΔA
ΔA - change in surface area

(c) Surface tension tends to compress the gas inside a **bubble**. Compression proceeds until the increase in internal pressure balances surface tension

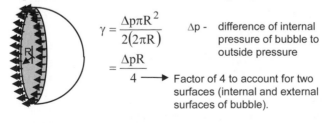

$\gamma = \dfrac{\Delta p \pi R^2}{2(2\pi R)}$ Δp - difference of internal pressure of bubble to outside pressure

$= \dfrac{\Delta p R}{4}$ → Factor of 4 to account for two surfaces (internal and external surfaces of bubble).

(d) Surface tension tends to compress the liquid inside a **drop**. Compression proceeds until increase in internal pressure balances surface tension.

$\gamma = \dfrac{\Delta p \pi R^2}{(2\pi R)}$

$= \dfrac{\Delta p R}{2}$ → A liquid drop has only one (outside) surface thus factor of 2 instead of 4

5.1.7 Capillary action

Consider a narrow tube placed in a container of liquid. It is observed that the level of liquid in the tube is different to that of the level in the container. The level in the tube may be higher or lower than the surrounding liquid. It depends on the contact angle. A narrow tube in which this occurs is called a **capillary**.

Contact angle $\theta < 90°$.
Surface is pulled up by surface tension forces F_R.

Contact angle $\theta > 90°$.
Surface is pulled down by surface tension forces F_R.

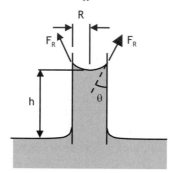
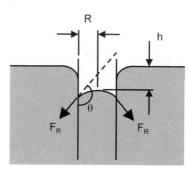

ρ - density of liquid
R - Radius of tube
θ - angle of contact (wetting angle)

Effects of **capillarity** are seen more clearly in small-scale apparatus since surface tension is a surface effect while weight is a volume effect. The coefficient of surface tension is given in this case by:

$$F_R \cos\theta = pA$$

Cos θ term is included since F_r acts at an angle to the weight of the liquid.

$$F_R = \frac{\rho gh}{\cos\theta} \pi R^2$$

$$l = 2\pi R$$

Length of contact is the circumference of the tube.

Thus:
$$\gamma = \frac{\rho gh}{\cos\theta} \frac{\pi R^2}{2\pi R}$$

$$= \frac{\rho ghR}{2\cos\theta}$$

5.1.8 Bernoulli's equation

In **laminar flow**, particles within a fluid move along smooth paths called **streamlines**. In a schematics diagram, the spacing of streamlines is an indication of the velocity of the fluid particles.

Consider the flow of fluid through a tube of varying cross-section:

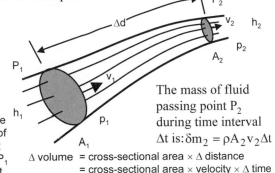

The mass of fluid passing point P_1 during time interval Δt is:

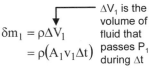

ΔV_1 is the volume of fluid that passes P_1 during Δt

$$\delta m_1 = \rho \Delta V_1$$
$$= \rho(A_1 v_1 \Delta t)$$

The mass of fluid passing point P_2 during time interval Δt is: $\delta m_2 = \rho A_2 v_2 \Delta t$

Δ volume = cross-sectional area × Δ distance
= cross-sectional area × velocity × Δ time

The fluid is incompressible ($\rho_1 = \rho_2$), and no fluid leaks out or is added through the walls of the pipe ($\delta m_1 = \delta m_2$). Thus:

$$\rho A_1 v_1 \Delta t = \rho A_2 v_2 \Delta t$$

The product Av is the volume flow rate Q in $m^3\ s^{-1}$. The volume flow rate is a constant for incompressible fluids.

$$\boxed{A_1 v_1 = A_2 v_2}$$ **Equation of continuity**

Consider a mass element $\delta m_1 = \rho A_1 v_1 \Delta t$ passing a point P_1 during a time interval Δt. Work done *on* the mass element δm_1

$$W = (p_1 A_1)\Delta d$$
$$= p_1 V_1$$
$$= p_1 \frac{\delta m}{\rho}$$

The same amount of mass $\delta m_2 = \delta m_1$ passes point P_2 during time interval Δt.

Work done *by* the mass element δm:

$$W = p_2 \frac{\delta m}{\rho}$$

Net mechanical work done *on* system:

$$\Delta W = \frac{\delta m}{\rho}(p_1 - p_2)$$

Change in kinetic energy of a mass element Δm: $\Delta KE = \frac{1}{2}\delta m(v_2^2 - v_1^2)$

Change in potential energy of a mass element Δm: $\Delta U = \delta mg(h_2 - h_1)$

Conservation of energy between P_1 and P_2:

$$\Delta W = \Delta PE + \Delta KE$$

$$\frac{\delta m}{\rho}(p_1 - p_2) = \delta mg(h_2 - h_1) + \frac{1}{2}\delta m(v_2^2 - v_1^2)$$

Absolute pressure

$$\boxed{p_1 + \rho g h_1 + \frac{1}{2}\rho v_1^2 = p_2 + \rho g h_2 + \frac{1}{2}\rho v_2^2}$$ **Bernoulli's equation**

5.1.9 Viscosity

Consider a fluid held between two plates. The bottom plate is fixed, and a force is applied to the top plate causing it to move.

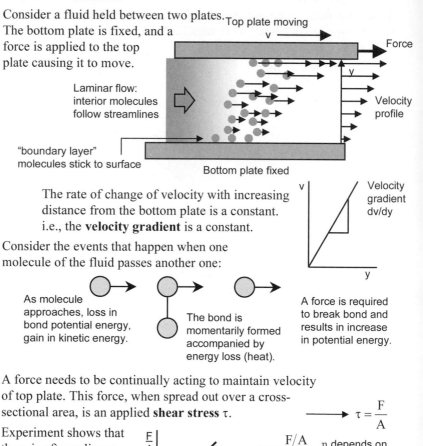

Laminar flow: interior molecules follow streamlines

"boundary layer" molecules stick to surface

The rate of change of velocity with increasing distance from the bottom plate is a constant. i.e., the **velocity gradient** is a constant.

Consider the events that happen when one molecule of the fluid passes another one:

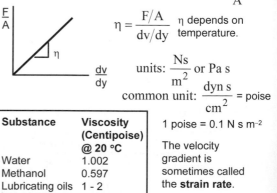

As molecule approaches, loss in bond potential energy, gain in kinetic energy.

The bond is momentarily formed accompanied by energy loss (heat).

A force is required to break bond and results in increase in potential energy.

A force needs to be continually acting to maintain velocity of top plate. This force, when spread out over a cross-sectional area, is an applied **shear stress** τ.

$$\tau = \frac{F}{A}$$

Experiment shows that there is often a linear relationship between the applied shear stress and the velocity gradient. The constant of proportionality (or slope) is called the **coefficient of viscosity** η. When η is a constant, the fluid is called a **Newtonian fluid**.

$$\eta = \frac{F/A}{dv/dy}$$

η depends on temperature.

units: $\dfrac{Ns}{m^2}$ or Pa s

common unit: $\dfrac{dyn\,s}{cm^2}$ = poise

1 poise = 0.1 N s m-2

The velocity gradient is sometimes called the **strain rate**.

Substance	Viscosity (Centipoise) @ 20 °C
Water	1.002
Methanol	0.597
Lubricating oils	1 - 2

5.1.10 Fluid flow

A spherical body moving through a viscous medium experiences a
resistive force which is proportional to its:

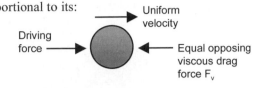

- velocity
- radius
- coefficient of
  viscosity of the
  medium

Driving force

Uniform velocity

Equal opposing viscous drag force F_v

Experiments show that the resistive force is given by:

$$F_v = 6\pi\eta vR$$ **Stokes' law**

- Stokes' law applies only for streamline conditions and when boundary
  layer molecules remain stationary on the surface of the sphere.

If a body falls through a viscous medium, it will
accelerate until the viscous resistive force and the
upthrust due to buoyancy equals the weight of the body.
It then reaches its **terminal velocity**.

This term increases with increasing velocity

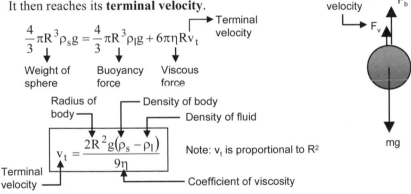

$$\frac{4}{3}\pi R^3 \rho_s g = \frac{4}{3}\pi R^3 \rho_l g + 6\pi\eta R v_t$$

Terminal velocity

Weight of sphere

Buoyancy force

Viscous force

Radius of body

Density of body

Density of fluid

$$v_t = \frac{2R^2 g(\rho_s - \rho_l)}{9\eta}$$ Note: v_t is proportional to R^2

Terminal velocity

Coefficient of viscosity

F_b

F_v

mg

For the flow of a fluid in a narrow tube, the volume passing through the
tube per second depends on:

- pressure difference Δp
- radius of tube (R)
- length of tube (l)
- viscosity of fluid η

$$Q = \frac{Vol}{t}$$

Assuming streamline flow: $$Q = \frac{\pi \Delta p R^4}{8\eta l}$$

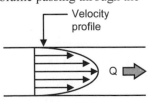

Velocity profile

Q

Note the strong dependence
on the tube radius (to the
fourth power).

5.1.11 Examples

1. A lead sphere of mass 5 kg is placed into a beaker of mercury at room
   temperature. Calculate the %volume of the sphere that floats above the
   surface of the mercury.

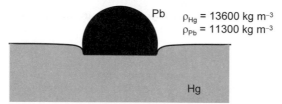

Pb ρ_{Hg} = 13600 kg m⁻³

ρ_{Pb} = 11300 kg m⁻³

Hg

Solution:

Let the volume of Hg displaced $= \Delta V$. Then the mass of fluid displaced is:

$$m_{Hg} = \rho_{Hg}\Delta V$$

and this is equal to the mass of the sphere (5 kg). Thus, if V is the total
volume of the sphere, then:

$$m_{Pb} = \rho_{Hg}\Delta V$$

$$= \rho_{Pb}V$$

$$\frac{\Delta V}{V} = \frac{\rho_{Pb}}{\rho_{Hg}} = \frac{11300}{13600} = 83\%$$

2. Transmission oil is pumped through a 10 mm diameter pipe 1.2 m long
   under a pressure difference of 200 kPa. What is the flow rate of oil
   (Litres/sec) if the oil is cold (η = 2 Pa.s) and then hot (η = 0.1 Pa.s)

Solution:

$$Q = \frac{\text{Vol}}{t}$$

$$Q = \frac{\pi \Delta p R^4}{8\eta l}$$

$$Q_{cold} = \frac{\pi 200 \times 10^3 \left(0.005^4\right)}{8(2)1.2}$$

$$= 20.4 \, mLs^{-1}$$

$$Q_{hot} = \frac{\pi 200 \times 10^3 \left(0.005^4\right)}{8(0.1)1.2}$$

$$= 409 \, mLs^{-1}$$

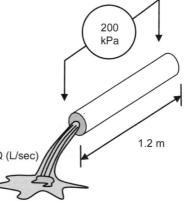

200 kPa

1.2 m

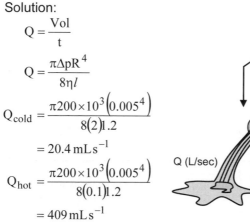

Q (L/sec)

5.2 Solids

Summary

$F = kx$	Hooke's law
$\sigma = \dfrac{F}{A}$	Stress
$\tau = \dfrac{F}{A}$	Shear stress
$\varepsilon = \dfrac{\Delta l}{l}$	Strain
$\varepsilon = \dfrac{\Delta V}{V}$	Volume strain
$B = \dfrac{\Delta \sigma_h}{\Delta V / V}$	Bulk modulus
$G = \dfrac{\tau}{\gamma}$	Shear modulus
$\nu = \dfrac{\dfrac{\Delta w}{w}}{\dfrac{\Delta l}{l}}$	Poisson's ratio
$W = \dfrac{1}{2} kx^2$	Strain energy
$\sigma = E\varepsilon$	Elastic modulus
$\begin{aligned} F &= kx + \lambda \dfrac{dx}{dt} \\ &= (k + i\omega\lambda)x \end{aligned}$	Viscoelasticity

5.2.1 Hooke's law

Robert Hooke (in 1676) found (by doing experiments) that if a certain force was needed to stretch a bar by x, then double the force was needed to stretch the same bar by 2x. Mathematically, this was expressed by Hooke as:

$$F = kx$$ **Hooke's law**

k depends on the type of material and the dimensions of the specimen.

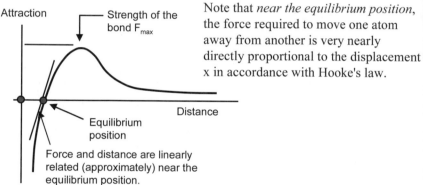

Consider the forces acting between two atoms:

Note that *near the equilibrium position*, the force required to move one atom away from another is very nearly directly proportional to the displacement x in accordance with Hooke's law.

Attraction — Strength of the bond F_{max}

Distance

Equilibrium position

Force and distance are linearly related (approximately) near the equilibrium position.

Repulsion

Thomas Young (in 1807) described Hooke's relationship in such a way which did not rely on the geometry of a particular specimen.

$$F = kx$$ Start with Hooke's law

$$\frac{F}{A} = \frac{kx}{A}$$ Divide both sides by A, the cross-sectional area of the specimen

$$= \frac{kl}{A}\frac{x}{l}$$ Multiply and divide by l, the length of the specimen

Let $E = \frac{kl}{A}$ E is a material property which describes the elasticity, or **stiffness** of a material and is called **Young's modulus**.

$$\frac{F}{A} = E\frac{x}{l}$$

Stress ⎵   ⎿ **Strain**

E_{steel}	210 GPa
$E_{Aluminium}$	70 GPa
E_{glass}	70 GPa

5.2.2 Stress

Tensile stress

When forces tend to pull on a body and thus stretch or elongate it, **tensile stresses** are produced *within the material*.

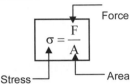

$$\sigma = \frac{F}{A}$$

Force

Stress ⟶

Area over which force acts

The units of stress are Pa (same as pressure).

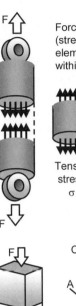

Force per unit area (stress) acts on an element of material within the body

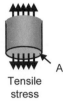

Tensile stress

σ

A

Compressive stress

When forces tend to push on a body and thus shorten or compress it, **compressive stresses** are produced within the material.

Compression and tension are called **normal stresses**.

↓

Because the force producing the stress acts **normal** to the planes under consideration. Symbol σ used for normal stresses.

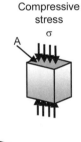

Compressive stress

σ

A

Shear stress

Force acting parallel to area produces **shear stress** τ.

$$\tau = \frac{F}{A}$$

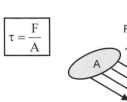

F

F

A

F

5.2.3 Strain

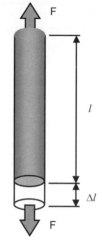

Linear strain

Application of a deforming force causes atoms within the body to be shifted away or *displaced* from their equilibrium positions. The net effect of this is a measurable change in dimensions of the body.

Strain is the fractional change in length of a body subjected to a deforming force.

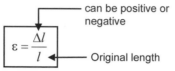

can be positive or negative

$$\varepsilon = \frac{\Delta l}{l}$$

Original length

Shear strain

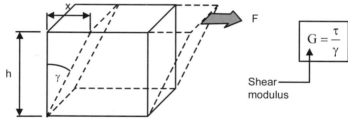

$$G = \frac{\tau}{\gamma}$$

Shear modulus

Shearing angle γ ⟶ Referred to as the
$\tan \gamma = x/h$ **shear strain**
or $\gamma = x/h$ for small deflections

Volume strain

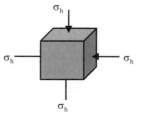

When a solid is subjected to uniform pressure over its whole surface, then the deformation is described by the **volume strain**.

$$\varepsilon = \frac{\Delta V}{V}$$

The **bulk modulus** is the ratio of the change in hydrostatic pressure over the volume strain

$$B = \frac{\Delta \sigma_h}{\Delta V / V}$$

5.2.4 Poisson's ratio

It is observed that for many materials, when stretched or compressed along the length within the elastic limit, there is a contraction or expansion of the sides as well as an extension or compression of the length. **Poisson's ratio** is the ratio of the fractional change in one dimension to the fractional change of the other dimension.

$$\nu = \frac{\dfrac{\Delta w}{w}}{\dfrac{\Delta l}{l}}$$

Poisson's ratio is a measure of how much a material tries to maintain a constant volume under compression or tension.

Consider a bar of square cross section w × w placed in tension under an applied force F. The initial total volume of the bar is:

$$V_1 = A_1 l$$

Where $A_1 = w^2$. After the application of load, the length of the bar increases by Δl. The width of the bar decreases by Δw. The volume of the bar is now calculated from:

$$V_2 = (l + \Delta l)(w - \Delta w)^2$$

$$= l(1+\varepsilon)w^2\left(1 - \frac{\Delta w}{w}\right)^2$$

$$= l(1+\varepsilon)A_1(1 - \nu\varepsilon)^2$$

$$\approx A_1 l(1 + \varepsilon - 2\nu\varepsilon) \quad \substack{\text{since} \\ \varepsilon^2 \ll 1}$$

The change in volume is thus:

$$V_2 - V_1 \approx A_1 l - A_1 l(1 + \varepsilon - 2\nu\varepsilon)$$

$$= A_1 l \varepsilon(1 - 2\nu)$$

For there to be no volume change, then ν has to be less than 0.5. $\nu > 0.5$ implies that the volume decreases with tension, an unlikely event. When $\nu = 0.5$, there is no volume change and the contraction in width is quite pronounced (e.g. rubber). When $\nu = 0$, the volume change is the largest and there is no perceptible contraction in width. Most materials have a value of ν within the range 0.2 to 0.4.

When the material contracts inwards (a so-called **plane stress** condition) under an applied tensile stress σ_T, there is no sideways stress induced in the material. If the sides of the material are held in position by external forces or restraints (**plane strain**), then there is a stress σ induced, the value of which is given by: $\sigma = \nu\sigma_T$

In terms of stresses and strains, in plane strain conditions (sides held in position), there is an effective increase in the stiffness of the specimen due to the induced sideways stresses. **Hooke's law** becomes:

$$\sigma = \frac{E}{1 - \nu^2}\varepsilon$$

5.2.5 Mechanical properties of materials

Hooke's law applies for the linear elastic region. When load is removed, body returns to its original shape. If body is stretched beyond the elastic limit, it will only partially return to its original shape and thus acquire a **permanent set**.

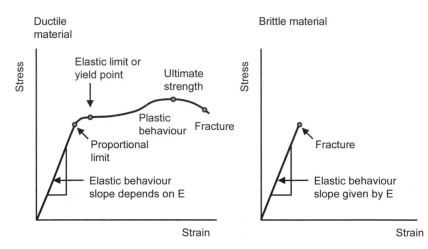

When a solid is stretched or compressed by the application of an applied force, the application of the force F through a distance dx requires work to be done on the system.

$$dW = Fdx$$

$$F = kx$$

$$W = \int kx\,dx$$

$$= \frac{1}{2}kx^2 \quad \text{Strain potential energy}$$

When a solid fractures, the stored **strain potential energy** is converted into heat, kinetic energy, plastic deformation and surfaces - that is, the surface energy of the cracked parts. **Brittle materials** generally shatter into many surfaces which soak up the stored strain energy released during fracture. **Ductile materials** tend to absorb the stored strain energy in the fracture surfaces and also **plastic deformation** inside the material.

5.2.6 Linear elasticity

Consider the shape of the force law between two atoms or molecules in more detail. Its shape resembles that of a sine wave in the vicinity of the force maximum.

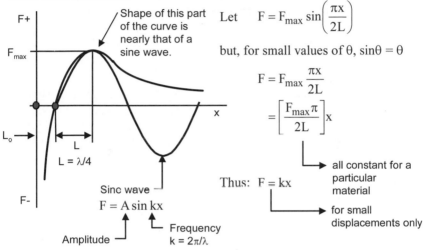

Shape of this part of the curve is nearly that of a sine wave.

Let $F = F_{max} \sin\left(\dfrac{\pi x}{2L}\right)$

but, for small values of θ, $\sin\theta = \theta$

$$F = F_{max} \frac{\pi x}{2L}$$

$$= \left[\frac{F_{max}\pi}{2L}\right]x$$

all constant for a particular material

Thus: $F = kx$

for small displacements only

Sinc wave
$F = A \sin kx$

Amplitude

Frequency
$k = 2\pi/\lambda$

F may be expressed in terms of force per unit area (or **stress**) which is given the symbol σ

$$\sigma = \frac{\sigma_{max}\,\pi}{2L}x$$

Let the fractional change in displacement from the equilibrium position (the **strain**) be given by:

$$\varepsilon = \frac{x}{L_o}$$

Substituting for x and transferring ε gives:

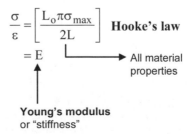

$$\frac{\sigma}{\varepsilon} = \left[\frac{L_o \pi \sigma_{max}}{2L}\right] \quad \textbf{Hooke's law}$$

$$= E$$

All material properties

Young's modulus
or "stiffness"

5.2.7 Viscoelastic materials

Most materials are neither completely elastic or completely fluid but fall within these two extremes. For materials which display an appreciable "fluid like" behaviour, even though they might appear to be solid, are called **viscoelastic**. Their mechanical properties are described in terms of an elastic modulus and a viscosity. The viscous component usually affects the response of a material when subjected to a changing force. The elastic properties usually are more important when materials are subjected to a static force.

The response of materials and systems can often be modelled by **springs** and **dashpots**. This allows both static and dynamic processes to be modelled mathematically with some convenience. Springs represent the solid-like characteristics of a system and is equivalent to the **elastic modulus** E. Dashpots represent the fluid-like aspects of a system and the **damping coefficient** λ is related to **viscosity** η.

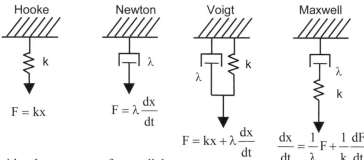

Consider the response of a parallel spring and dashpot to an sinusoidal force. The resulting displacement x is also sinusoidal but is out of phase with the force.

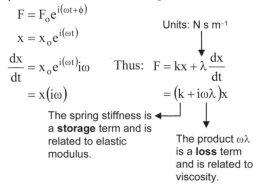

$$F = F_0 e^{i(\omega t + \phi)}$$

$$x = x_0 e^{i(\omega t)}$$

$$\frac{dx}{dt} = x_0 e^{i(\omega t)} i\omega$$

$$= x(i\omega)$$

The spring stiffness is a **storage** term and is related to elastic modulus.

Units: N s m⁻¹

Thus: $F = kx + \lambda \dfrac{dx}{dt}$

$$= (k + i\omega\lambda)x$$

The product $\omega\lambda$ is a **loss** term and is related to viscosity.

The damping coefficient λ has the same relationship to viscosity η as the spring stiffness k does to the elastic modulus E. The damping coefficient and stiffness apply to a particular specimen geometry where the modulus and viscosity apply to the material (i.e. E and η are material properties).

5.2.8 Lattice waves

When a solid absorbs heat energy, its temperature rises. The energy being absorbed is converted to **internal energy**. In solids, molecules do not translate or rotate (otherwise the material wouldn't be solid!). Internal energy is the **vibrational energy** of the constituent atoms.

Above absolute zero of temperature, atoms are vibrating around their equilibrium positions with a very high frequency $\approx 10^{14}$Hz.

These high frequency vibrations travel throughout the solid as **elastic waves** with a velocity equal to the speed of sound in the material.

Internal energy is the energy carried by the elastic lattice waves within the material. An increase in internal energy resulting from a change in temperature corresponds to an increase in the amplitude of the lattice waves.

Lattice waves in a solid

In **classical thermodynamics**, we might treat each vibrating atom as a **harmonic oscillator** (i.e. the atoms are connected to other atoms by linear springs undergoing **simple harmonic motion**). For a single atom, there may be oscillations in three directions and thus, the energy of vibration is:

$$U = 3kT$$

For a mole of atoms, the total energy is:

$$U = 3N_A kT$$

$$= 3RT \quad \text{Since } R = N_A k$$

$$R = 8.3145 \text{ J mol}^{-1} \text{ K}^{-1}$$

But, $\dfrac{dU}{dT} = C$ the molar specific heat.

Thus: $\boxed{C = 3R}$ **Dulong-Petit law**

The energy of vibration of an atom (PE to KE to PE etc) is the same no matter what the molecular or atomic mass. The total energy depends on the total NUMBER of atoms. The (mass) specific heat of solids with a low molecular weight is larger than that for solids with a high molecular weight because in the former, there are a greater number of molecules in 1kg of material than there are in the latter.

That is, the **molar specific heat** for all solids is a constant equal to 3R. This is true for most solids at reasonably high temperatures but is not observed to hold at low temperatures close to absolute zero.

5.2.9 Phonons

Matter is not continuous. The regular spacing of an atomic lattice introduces a discreteness in the allowable frequencies, and hence energies, of lattice waves.

Just as the energy of electromagnetic waves is carried by **photons**, the allowable energies associated with lattice waves are called **phonons**.

When the energies of phonons are calculated, it can be shown that the molar specific heat changes with decreasing temperature and is in agreement with experimental observations.

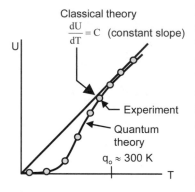

Classical theory
$$\frac{dU}{dT} = C \text{ (constant slope)}$$
Experiment
Quantum theory
$q_o \approx 300$ K

Phonons are important because internal energy is the energy of the phonons within a solid. In metals, the **conduction electrons** increase their kinetic energy with increasing temperature and thus contribute to the specific heat. But the major proportion of internal energy occurs as phonon energy.

Quantum theory agrees with classical theory above the **Debye temperature** q_D but agrees with experimental observations at low temperatures.

Thermal energy is transported from hot to cold regions within a solid by both phonons and free electrons. The total thermal conductivity k is the summation of these conductivities : $k = k_p + k_e$

Heat conduction in **insulators** is entirely via phonons. Anything which inhibits the travel of, or scatters, phonons leads to a decrease in thermal conductivity. Lattice imperfections scatter phonons and thus **amorphous** substances (glass) have a lower thermal conductivity than **crystalline** materials. Increasing the temperature also results in increasing scattering thus k usually decreases with increasing temperature.

Thermal conductivity in **conductors** (e.g. metals) arises almost entirely from the transport of energy by free electrons. Good thermal conductors are also good electrical conductors because of the large number density of free electrons. The two conductivities are physically related:

$$L = \frac{k}{\sigma T}$$

L is called the **Lorentz number** and is a constant.

Material	k (W m⁻¹ K⁻¹)
Aluminium	220
Steel	54
Glass	0.79
Water	0.65
Glass wool	0.037
Air	0.034

5.2.10 Examples

1. A cube of aluminium is subjected to a hydrostatic pressure of 4GPa. Determine the % change in length of the side of the cube.

   $B_{Al} = 7.46 \times 10^{10}$ GPa

Solution:

$$B = \frac{\Delta\sigma_h}{\Delta V/V} \qquad\qquad V = l^3$$

$$7.46 \times 10^{10} = \frac{4 \times 10^9}{\Delta V/V} \qquad \frac{\Delta V}{V} = 3\frac{\Delta l}{l}$$

$$\frac{\Delta V}{V} = 5.3\% \qquad\qquad 5.3 = 3\frac{\Delta l}{l}$$

$$\frac{\Delta l}{l} = 1.78\%$$

2. A 60mm cube of copper is subjected to a shearing force of 150 kN. The top face of the cube is displaced 0.25 mm with respect to the bottom. Calculate:

   (a) the shearing stress
   (b) the shearing strain
   (c) the shear modulus

Solution:

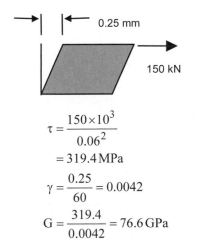

0.25 mm

150 kN

$$\tau = \frac{150 \times 10^3}{0.06^2}$$
$$= 319.4\,\text{MPa}$$

$$\gamma = \frac{0.25}{60} = 0.0042$$

$$G = \frac{319.4}{0.0042} = 76.6\,\text{GPa}$$

5.3 Matter

Summary

$$E = hf$$ Photon energy

$$\lambda_{max} = \frac{hc}{4.97k}$$ Wien's displacement law

$$\frac{1}{2}mv^2 = hf - W$$ Photoelectric effect

$$\frac{1}{\lambda} = R\left(\frac{1}{2^2} - \frac{1}{n^2}\right)$$ Balmer formula

$$\lambda = \frac{h}{mv}$$ de Broglie matter waves

$$\Delta p \Delta x \geq \frac{h}{2\pi}$$ Heisenberg uncertainty principle

$$-\frac{\hbar}{2m}\frac{\delta^2 \psi}{\delta x^2} + V\psi = i\hbar\frac{\delta \psi}{\delta t}$$ Schroedinger equation

$$E_T = mc^2 = E_K + m_o c^2$$ Einstein's equation

5.3.1 Radiation emission spectrum

Experiments show that the distribution of intensity of radiation with wavelength emitted from a black body has a characteristic shape which depends upon the body's temperature.

Maxwell calculated the velocity distribution of molecules in an ideal gas and found that the distribution had a characteristic shape that also depended on the temperature of the gas.

Wien argued that since the molecules within a heated body are vibrating with thermal energy, then the Maxwell distribution of velocities would result in acceleration of charges within the molecules thus leading to the emission of radiation with a characteristic intensity spectrum. Wien's predictions agreed well with experimental results at high frequencies but did not fit well at low frequencies. Rayleigh and Jeans performed more rigorous calculations

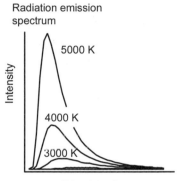

Radiation emission spectrum

and found that the predicted radiation emission spectrum agreed well at low frequencies but had an ever increasing intensity at higher frequencies - a feature they termed the ultraviolet catastrophe.

In 1901, Planck found that the emission spectrum could only be explained in terms of Maxwell's statistics and Boltzmann's statistical interpretation of entropy for the radiation emitted by electric oscillators (little was known about the structure of the atom at this point) if the radiation emission was allowed to only occur in discrete amounts which he called energy quanta. Planck determined that the energy distribution could be expressed as:

$$\psi(\lambda) = \frac{8\pi}{\lambda^4} \frac{E}{e^{E/kT} - 1}$$

k is Boltzmann's constant: 1.38×10^{-23} J K^{-1}

Hz

as long as the energy term E had a minimum value given by: $\boxed{E = hf}$

The maximum in the emission spectrum is found by differentiating Planck's energy distribution with respect to λ. This yields:

Planck's constant: 6.626×10^{-34} J s^{-1}

$$\boxed{\lambda_{max} = \frac{hc}{4.97k}}$$ **Wien's displacement law**

5.3.2 Photoelectric effect

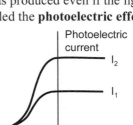

In **thermoionic emission**, electrons can be ejected from a hot filament as a result of the kinetic energy imparted to them. These electrons can form an electric current if an external field is applied.

In 1887, **Hertz** observed that a current could be created if a metal was illuminated by light of a sufficiently high frequency. Experiments showed that the current in this case could only be produced if the frequency of the light was above a critical value (the **threshold frequency**). If the frequency was below this value, no current was produced even if the light was of very high intensity. This effect was called the **photoelectric effect** and for many years, remained unexplained.

When electrons are ejected, it is found that even with no applied potential at the cathode, there is still a very small current. A small reverse voltage (the **stopping potential**) is needed to stop all the **photoelectric current**.

The explanation of the photoelectric effect was given by **Einstein** in 1905. Einstein postulated that light consisted of energy **quanta** in accordance with Planck's equation. If the energy of the incoming light was greater than the work function of the metal surface, then the excess energy would be imparted to the electron as kinetic energy. The maximum kinetic energy is given by:

$$\frac{1}{2}mv^2 = hf - W$$

$$= q_e V_o$$

Charge on electron
1.602×10^{-19} C

The stopping potential is a measure of the maximum kinetic energy of the ejected electrons. If the stopping potential is plotted against frequency, the slope of the resulting linear function is Planck's constant and the intercept is the work function.

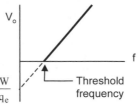

The work function is a measure of the surface energy potential. It is on the order of a few eV for most metals.

5.3.3 Line spectra

Ever since the 18th century, it was known that the **emission spectrum** from a heated gas consisted of a series of lines instead of a continuous rainbow of colours. The position (or wavelength) of spectral lines was known to be unique to each type of element.

When white light is shone through a cool gas, it is found that dark lines appear in the resulting spectrum, the position of which correspond exactly with the bright line spectra obtained when the gas is heated.

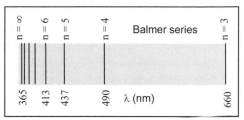

Physicists in the 19th century used emission and **absorption spectra** to identify gases in the atmosphere of the sun. In 1885, **Balmer** formulated an empirical equation that accounted for the position of the lines in the visible part of the hydrogen spectrum.

$$\frac{1}{\lambda} = R\left(\frac{1}{2^2} - \frac{1}{n^2}\right)$$

The Balmer formula applies only to the hydrogen atom. **Rydberg** proposed a more general formula for the heavier elements.

Rydberg constant: 1.0973731×10^7 m^{-1}

In this formula, n is the line number and takes on the values 3, 4, 5, …The formula predicts an infinite number of spectral lines which become closer together with increasing value of n. At $n = \infty$, the wavelength is 364.6 nm, the limit of the **Balmer series**. Other series were since discovered in the hydrogen spectrum by letting the first term in the brackets equal 1, 3, 4, etc.

$$\frac{1}{\lambda} = R\left(\frac{1}{1^2} - \frac{1}{n^2}\right)$$ Lyman series (ultraviolet) n = 2, 3, 4...

$$\frac{1}{\lambda} = R\left(\frac{1}{2^2} - \frac{1}{n^2}\right)$$ Balmer series (visible) n = 3, 4, 5...

$$\frac{1}{\lambda} = R\left(\frac{1}{3^2} - \frac{1}{n^2}\right)$$ Paschen series (infrared) n = 4, 5, 6...

$$\frac{1}{\lambda} = R\left(\frac{1}{4^2} - \frac{1}{n^2}\right)$$ Brackett series (infrared) n = 5, 6, 7....

The existence of spectral lines could not be explained by classical physics. Balmer's equation demonstrated an order, and an integral order at that, to the position of lines within the frequency spectrum. Balmer did not propose any physical explanation for his formula, but simply noted that it described the phenomena almost exactly. The importance of the equation is that is provides a test for what was to become a model for atomic structure.

5.3.4 Bohr atom

In 1897, **Thomson** demonstrated that cathode rays (observed to be emitted from the cathodes of vacuum tubes) were in fact charged particles which he called electrons. Thomson proposed that the atom consisted of a positively charged sphere in which were embedded negatively charged electrons.

Rutherford subsequently found in 1911 that the electrons orbited at some distance from a central positively charged **nucleus**. Rutherford proposed that electrostatic attraction between the nucleus and the electron was balanced by the centrifugal force arising from the orbital motion. However, if this were the case, then the electrons (being accelerated inwards towards the centre of rotation) would continuously radiate all their energy as **electromagnetic waves** and very quickly fall into the nucleus.

In 1913, **Bohr** postulated two important additions to Rutherford's theory of atomic structure:

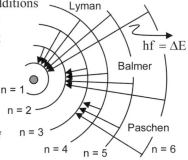

- Electrons can orbit the nucleus in what are called **stationary states** in which no emission of radiation occurs and in which the **angular momentum** is constrained to have values:

$$L = m_e vr = \frac{nh}{2\pi}$$

The 2π appears because L is expressed in terms of ω rather than f.

- Electrons can make transitions from one state to another accompanied by the emission or absorption of a single **photon** of energy $E = hf$ thus leading to absorption and emission spectra.

As in the Rutherford atom, the centrifugal force is balanced by Coulomb attraction:

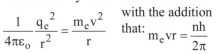

Mechanical model of hydrogen atom

$$\frac{1}{4\pi\varepsilon_o}\frac{q_e^2}{r^2} = \frac{m_e v^2}{r}$$ with the addition that: $m_e vr = \frac{nh}{2\pi}$

By summing the kinetic energy (from the orbital velocity) and the potential energy from the electrostatic force, the total energy of an electron at a given energy level n is given by:

$$E_n = -\frac{m_e Z^2 q_e^4}{8\varepsilon_o^2 h^2 n^2}$$

Note, Z = 1 for the hydrogen atom where the energy of the ground state is 13.6 eV. The energy levels for each state n rises as Z^2, thus, the energy level of the innermost shell for multi-electron atoms can be several thousand eV.

from which the Rydberg constant may be calculated since $\Delta E = hf$

5.3.5 Energy levels

The stationery states or energy levels allowed by the Bohr model of the atom are observed to consist of sub-levels (evidenced by fine splitting of spectral lines). These groups of sub-levels are conveniently called **electron shells**, and are numbered K, L, M, N etc, with K being the innermost shell corresponding to n = 1. The number n is called the **principle quantum number** and describes how energy is quantised.

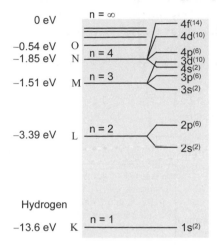

The energy required to move an electron from an electron shell to infinity is called the **ionisation energy**. It is convenient to assign the energy at infinity as being 0 since as an electron moves closer to the nucleus (which is positively charged) its potential to do work is less, thus the energy levels for each shell shown are negative. For hydrogen, the ionisation energy is -13.6 eV. The energies for the higher energy levels is given by:

$$E = -\frac{13.6}{n^2} \quad \text{For Hydrogen}$$

The electron-volt is a unit of energy.
1 eV = 1.609 × 10⁻²³J

At each value of n (i.e. at each energy level) the **angular momentum** can take on several distinct values. The number of values is described by a second quantum number l. The allowed values of l are 0, 1, … (n−1). Each value of l is indicated by a letter: ⟶

A third quantum number m describes the allowable changes in angle of the **angular momentum** vector in the presence of an electric field. It takes the values $-l$ to 0 to $+l$.

A fourth quantum number describes the **spin** of an electron where the spin can be either −1/2 or +1/2.

$l = 0$	s
$l = 1$	p
$l = 2$	d
$l = 3$	f
$l = 4$	g
$l = 5$	h

According to the **Pauli exclusion principle**, no electron in any one atom can have the same combination of quantum numbers. This provides the basis for the filling of energy levels.

When all the electrons in an atom are in the lowest possible energy levels, the atom is said to be in its **ground state**.

For example the 3d energy level can hold up to 10 electrons:

n = 3

thus: l = 0, 1, 2 = s, p, or d

and: m = -2, -1, 0, 1, 2

5 values of m times two for spin thus 10 possible electrons

5.3.6 X rays

X rays are produced by electrons being decelerated from an initial high speed by collisions with a target material. Electrons may be produced by **thermoionic emission** and given a high velocity by the application of an electric field.

When an electron (q_e) collides with a target material, it is rapidly decelerated and a photon is emitted. The wavelengths of the photons involved are mainly in the X ray region of the electromagnetic spectrum. The most rapid decelerations result in the shortest wavelength photons. For other collisions, the electron may lose energy as photons of larger wavelength and also to heat by increasing the vibrational internal energy of the target. The result is a continuous spectrum of photon energies with a minimum wavelength dependent upon the kinetic energy of the electrons.

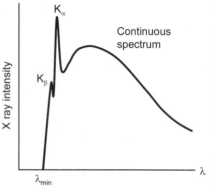

When an X ray is incident on a material, electrons within the target in the K and L shells can be ejected. Since there are relatively few electrons in these innermost shells, X rays do not have a high probability of interacting with them and hence their high penetrating ability.

Incoming electrons may also ionise the atoms of the target by ejecting bound electrons from within material. Some of these ejected electrons may come from the innermost energy levels which, in solid, can have energies in excess of 100,000 eV. An outer electron can fall into this vacancy and emit a photon in the process.

X rays resulting from filling of K shell vacancies by an electron from the L shell are called K_α X rays. X rays from M to K shell transitions are K_β, and those from N (and higher) to K transitions are K_γ. Similarly, transitions from M to L are L_α, N (and higher) are L_β.

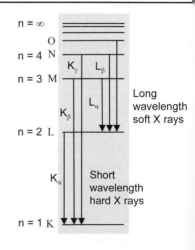

5.3.7 Matter waves

The Bohr model of the atom strictly applies only to a single electron orbiting a nucleus and ignores interactions between electrons and other neighbouring atoms. Further, the theory does not offer any explanation as to *why* the angular momentum is to be quantised. Such explanations and treatments can only be explained in terms of **wave mechanics**.

In 1924 **de Broglie** postulated that matter exhibited a dual nature (just as did electromagnetic radiation) and proposed that the wavelength of a particular object of mass m is found from:

$$\lambda = \frac{h}{mv}$$
Because h is a very small number, the wavelength of large objects is very small. For small objects, e.g. electrons, the wavelength is comparable to atomic dimensions.

where mv is the momentum p of the object. The resulting waves are called **matter waves**. In the case of atomic structure, matter waves for electrons are **standing waves** that correspond to particular electron orbits.

For a particular radius r, a standing wave is obtained when the circumference of the path is an integral number of wavelengths: $n\lambda = 2\pi r$

Thus, from the expression for matter waves, we obtain:
$$2\pi r = n\left(\frac{h}{mv}\right)$$

$$mvr = n\frac{h}{2\pi} \longrightarrow \text{Bohr condition for stable state since L = mvr.}$$

The **wave-particle duality** of matter means that inherently, an electron is neither a wave or a particle but its motion can be quantified using the mathematical equations appropriate to waves and particles. The wave nature of matter is often interpreted as being one of probabilities. The amplitude of a matter wave represents the probability of finding the associated particle at a particular position.

Since matter is described in terms of a probability, there becomes an inherent limitation in what we can know about the motion and position of a particle such as an electron. The **Heisenberg uncertainty principle** quantifies these uncertainties. For momentum and position, the requirement is:

$$\Delta p \Delta x \geq \frac{h}{2\pi}$$

Where Δp and Δx are the uncertainties associated with these quantities. The more we reduce the uncertainty in one, the more the uncertainty in the other increases.

5.3.8 Quantum mechanics

The total energy of a system is the sum of the potential and kinetic energies. Expressed in terms of momentum, p, this is stated:

$$E = \frac{p^2}{2m} + V$$

The value of the potential function may depend on both position and time. The form of V(x,t) is different for different arrangements of atoms (e.g. a single isolated atom, an atom in a regular array of a crystal).

Thus: $hf = \frac{p^2}{2m} + V(x,t)$

since $E = hf$

Let $p = -i\hbar \frac{\delta\Psi}{\delta x}$

$E = i\hbar \frac{\delta\Psi}{\delta t}$

Ψ is a variable, the form and value of which provides information about the motion of a wave/particle.

Thus:

$$-\frac{\hbar^2}{2m}\frac{\delta^2\Psi}{\delta x^2} + V(x,t)\Psi = i\hbar\frac{\delta\Psi}{\delta t}$$

Schroedinger equation

The solution to the wave equation is the **wave function** Ψ. If V is a function of x only, then the wave equation can be separated into time-independent and time-dependent equations that can be readily solved.

$$-\frac{\hbar^2}{2m}\frac{\delta^2\psi}{\delta x^2} + V(\psi) = E\psi \qquad \phi(t) = e^{i\frac{E}{\hbar}t}$$

The resulting solutions of these equations, when multiplied together, give the **wave function**: $\Psi(x,t) = \psi(x)\phi(t)$

The wave function gives all the information about the motion of a particle, such as an electron in an atom. Ψ is a complex quantity, the magnitude of which $|\Psi|$ is interpreted as a **probability density function** which in turn can be used to determine the probability of an electron being at some position between x and Δx.

Quantum mechanics is concerned with determining the wave function (i.e. solving the Schroedinger equation) for particular potential energy functions such as those inside atoms. It is found that valid solutions to the time-independent wave equation occur only when the energy is quantised. The solutions correspond to **stationary states**.

Solutions to the Schroedinger equation can be found for potential functions which are a function of both x and t. This enables time-dependent phenomena (e.g. the probability of transitions between energy levels in an atom) to be calculated and hence the intensity of spectral lines.

5.3.9 Radioactivity

Radioactivity was discovered by accident in 1896 by **Becquerel** who found that the rays emitted by a sample of uranium were similar to X rays (discovered a few months before by **Roentgen**) but appeared spontaneously from the uranium sample. These new rays had a very high penetrating power and could also cause **ionisation** of the air just like X rays. Soon after, many other naturally occurring **radioactive** elements were discovered, chief among these being radium and polonium by Marie **Curie** in 1911.

Rutherford found that a sample of uranium emitted two kinds of rays which he called **alpha** (α) and **beta** (β) rays. In 1900, **Villard** found a third type of ray from radium which were called **gamma** (γ) rays. One of the most striking properties of these rays was their penetrating power into matter. Experiments showed that alpha rays were the least penetrating but produced the greatest ionisation and gamma rays were the most penetrating but caused the least amount of ionisation.

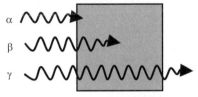

Alpha particles are the least penetrating because all their energy goes into ionisation of the material with which they interact. Gamma rays are the least ionizing and hence retain their energy and are the most penetrating.

Experiments in the early 1900's showed that a magnetic field could deflect α and β rays but not γ rays. Since the deflection of β rays was in the same direction as that observed by electrons in a magnetic field, it was concluded that β rays were in fact negatively charged electrons. Since the deflection of α rays was in the opposite direction, it was concluded that α rays were positively charged particles. Further experiments by Rutherford showed that α particles were **helium nuclei** consisting of two protons and two neutrons.

In 1913, experiments with **radioactive** materials indicated that the nucleus consists of positively charged protons and negatively charged electrons (i.e. electrons in addition to orbiting electrons). In 1932, **Chadwick** showed that the nucleus contains additional particles called **neutrons** which had a mass similar to that of a proton but no electric charge. The emission of a beta particle (an electron) from nucleus was thought to result from the *transformation* (rather than a separation) of a neutron to a proton $+\beta$. Experiments showed that to satisfy the law of conservation of energy, an additional particle was also emitted along with the beta particle from the nucleus. This particle was called the **neutrino**, it has a very small mass and carries no electric charge.

5.3.10 Half life

Radioactive decay is a random process. The probability that a particular radioactive nucleus (**radionuclide**) will decay in any selected time period is independent of the state of neighbouring nucleii, the chemical state, pressure and temperature. If there are N radioactive nucleii present, then the number that decay per unit time period (say a second) is given by:

$$\frac{dN}{dt} = -\lambda N \quad \longrightarrow \text{Units: sec}^{-1}$$

λ is a constant called the **decay constant**, which is different for different types of atoms. Notice that the decay rate depends on the total number of nucleii, or atoms, present. The decay constant indicates the probability that a single atom will decay in a unit of time. The larger the value of λ, the more rapidly the decay. Integrating the above equation, we obtain an expression for the number of remaining atoms present after a time t:

$$N_t = N_0 e^{-\lambda t} \quad \text{At t = 0, } N_t = N_o$$

$$\ln N_t = -\lambda t + \ln N_o$$

The decay constant can be determined by experiment. The number of disintegrations per second (dN/dt) can be measured with a **scintillation detector** or a **geiger counter** (depending on the nature of the material being studied). Thus, since dN/dt is proportional to N_t, a plot of ln(dN/dt) has a slope $-\lambda$.

$$N = N_0 - N_t$$

$$\frac{dN}{dt} = -\lambda N$$

$$= -\lambda(N_0 - N_t)$$

$$= \frac{dN_t}{dt} - \lambda N_0$$

$$\ln\left(\frac{dN}{dt}\right) \propto \ln N_t \quad \text{constant}$$

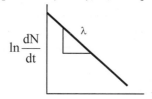

$$\ln \frac{dN}{dt}$$

A convenient measure of radioactive decay is the half-life $t_{1/2}$. This is the time for one half of the radioactive nucleii present to decay.

$$\frac{N_t}{N_o} = e^{-\lambda t_{1/2}} = \frac{1}{2}$$

$$\ln \frac{1}{2} = -\lambda t_{1/2}$$

U^{238}	4.51×10^9 years
C^{14}	5568 years
Ba^{137}	2.64 mins

$$\boxed{t_{1/2} = \frac{0.69}{\lambda}}$$

5.3.11 Nuclear energy

By the 1930's it was known that **nuclear reactions** whether induced by naturally occurring radioactivity or artificially by bombarding elements with α, β particles, γ rays or neutrons result in a change in the chemical nature of the element being studied. For example, when gamma rays are used to bombard atoms of mercury, the resulting nuclear reaction occurs:

$$\gamma + {}_{80}Hg^{198} \rightarrow {}_{79}Au^{197} + {}_{1}H^{1}$$
γ ray   mercury   gold   proton

The dream of alchemists of the middle ages.

Nuclear reactions may absorb or release energy, but the energy involved is far greater than that associated with chemical reactions involving the breakage and formation of chemical bonds. In nuclear reactions, changes in mass result in large changes of energy in accordance with **Einstein**'s equation $E = mc^2$.

Mass number - relative mass of an atom with respect to 1/12 the mass of a carbon 12 atom.

It is found that the sum of the masses of individual protons, neutrons and electrons is always more than the mass of a complete atom. Consider a Carbon 12 atom ${}_6C^{12}$

Chemical symbol

$$\rightarrow A X^Z$$

Atomic number - number of protons on the nucleus.

Mass of 6 protons & 6 electrons	= 6.04695
mass of 6 neutrons	= 6.05199
Total mass of parts	= 12.09894
Mass of C^{12} atom	= 12.00000
Difference in mass	= 0.09894

This difference in mass represents an energy loss of 92.1 MeV when the atom was formed. This energy is called the **nuclear binding energy**. $1eV = 1.6 \times 10^{-19}$ J

In nuclear reactions, if the total **binding energy** of the products is greater than that of the reactants, then energy is liberated. That is, the mass of the products is less than the mass of the reactants. The distribution of the average binding energy of all the elements show that when two light nuclei combine to form a heavier nucleus (**fusion**), or when a heavy nucleus splits to form two lighter nuclei (**fission**), energy is released.

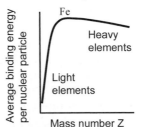

In stars, temperature is so high that hydrogen nuclei (positively charged protons) have enough kinetic energy to overcome their mutual repulsion and undergo fusion into helium nuclei releasing energy.

5.3.12 Nuclear physics

In the early 20th century it was believed that all matter consisted of protons, neutrons, electrons and neutrinos. Today, it is accepted that **quarks**, electrons and **messenger particles** responsible for the fundamental forces of nature are the basic building blocks of matter. Quarks and electrons are ordinary particles which contain **matter**. There are a number of different types of quark:

- Up
- Down
- Strange
- Charmed
- Bottom
- Top

Each type of quark comes in three different "colours": red, green and blue.

The names given to the types of quark are arbitrary, and simply serve as labels to distinguish one type of quark from another.

Nuclear particles (protons and neutrons) consist of three quarks. A proton consists of two up quarks and one down. A neutron consists of two down and one up.

The four fundamental forces of nature (**gravitational force, electromagnetic force**, the **weak nuclear force** and the **strong nuclear force**) can be thought of acting between two particles by the exchange of messenger particles.

In the nucleus, quarks are held together by the strong nuclear force by the exchange of particles called gluons. Electromagnetic attraction is caused by the exchange of particles called **photons**. Gravity is thought of as occurring by the exchange of **gravitons**.

Quantum theory shows how everything in the universe, including electromagnetic waves and gravity, consist of particles. These particles have a property called **spin**.

Particles with spin 1/2 are those that make up the matter of the universe (quarks and electrons) whereas particles with spin 0, 1 and 2 are the messenger particles which result in the fundamental forces between them. The property of spin lead **Dirac** to propose the existence of **antimatter**.

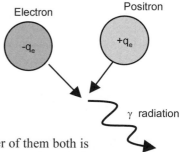

Electron

Positron

$-q_e$

$+q_e$

γ radiation

When antimatter and matter meet, the matter of them both is converted into energy in the form of electromagnetic radiation.

5.3.13 Example

1. The inwards force acting on an orbiting electron in the Bohr model of the atom arises from electro static attraction and is given by the Coulomb force law:

$$F = \frac{1}{4\pi\varepsilon_o} \frac{q_e^2}{r^2}$$

This force is balanced by the centripetal force given by:

$$F = \frac{m_e v^2}{r}$$

If, as Bohr postulated, that the angular momentum $L = mvr$ can only take on values such that:

$$m_e vr = n\frac{h}{2\pi}$$

determine an expression for the radius of the hydrogen atom when the electron is in the ground state ($n = 1$) and calculate this radius given the values of m_e and q_e below.

$m_e = 9.11 \times 10^{-31}$ kg
$q_e = 1.602 \times 10^{-19}$ C
$h = 6.626 \times 10^{-34}$ J s^{-1}

Solution:

Coulomb force = Centripetal force

$$\frac{1}{4\pi\varepsilon_o} \frac{Zq_e^2}{r^2} = \frac{m_e v^2}{r}$$

Hydrogen atom

$$v = \frac{nh}{2\pi m_e r} \quad \text{From Bohr condition}$$

$$\frac{1}{4\pi\varepsilon_o} \frac{Zq_e^2}{r^2} = \frac{m_e}{r}\left(\frac{nh}{2\pi m_e r}\right)^2$$

$$r = \frac{\varepsilon_o h^2 n^2}{\pi m_e Zq_e^2} \quad \text{Radius of electron orbit for hydrogen (Z = 1)}$$

$$= \frac{8.85 \times 10^{-12}\left(6.626 \times 10^{-34}\right)^2}{\pi\left(9.11 \times 10^{-31}\right)\left(1.6 \times 10^{-19}\right)^2}$$

$$= 5.298 \times 10^{-11}\,\text{m}$$

$$= 0.053\,\text{nm}$$

5.4 Universe

Summary

$$I = \frac{L}{4\pi d^2} \qquad \text{Intensity}$$

$$M = 4.75 - 2.5\log\frac{L}{L_o} \qquad \text{Magnitude}$$

$$4_1H^1 \rightarrow_2 HE^4 + 2_{+1}e^0 + 24.7\,MeV \qquad \text{Nuclear fusion}$$

$$F = G\frac{m_1 m_2}{d^2} \qquad \text{Gravitational force}$$

$$v = \sqrt{\frac{Gm_E}{R}} \qquad \text{Orbital velocity}$$

5.4.1 Copernicus, Tycho and Kepler

For **Aristotle**, the universe was a series of interconnecting spheres, the outermost containing the the stars which was moved by God, and by which motion was transmitted to the planets and then to the Moon, the innermost sphere. At the centre of the universe was the Earth. Mechanical details of accounting for the motion of planets was perfected by **Ptolemy** (150AD) through the use of **epicyles**, **equants** and **eccentrics** the combination of which explained the motion of heavenly bodies with uniform circular motion.

Copernicus (1473-1543) believed that the discrepancies between the observed motions of the planets and those predicted by the **geocentric** model of **Ptolemy** could be explained by a heliocentric model (proposed by some early Greek philosophers) but still containing epicycles and equants (indeed even more than Ptolemy) along with uniform circular motion.

Kepler (1571-1630), using the most accurate astronomical observations of **Tycho** Brahe, discovered that despite the attractiveness of uniform circular motion, it simply just did not fit the motion of the planets as observed.

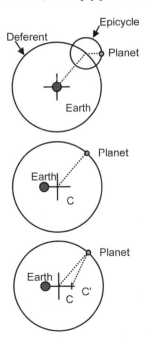

Epicycle

The planet moves along a small circle called an **epicycle**. The centre of the epicycle rotates about the Earth.

Eccentric

The planet moves along a circular path whose centre is offset from the position of the Earth giving it an **eccentric** orbit.

Equant

The planet moves along a circular path whose centre is offset from the position of the Earth. However, its velocity was a constant with respect to another point C' which is as far off centre as the Earth from C but in the other direction.

5.4.2 Kepler's laws

On the basis of **Tycho**'s observations, **Kepler** dispensed with the concept of uniform circular motion and discovered that:

- The orbit of a planets is an **ellipse** where the sun is at one focus
- the velocity of a planet along the orbit varies so that a line joining the sun to the planet sweeps out equal areas in equal intervals of time
- the square of the period of the orbit of a planet is proportional to the cube of planet's distance from the sun.

With these three **laws of planetary motion**, Kepler dispensed with epicycles, equants, eccentrics and a host of other detail - and indeed dispensed with the long held requirement for uniform circular motion.

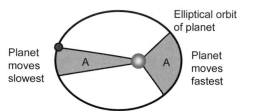

Elliptical orbit of planet

Planet moves slowest

Planet moves fastest

For any interval in time, no matter where on the orbit, the planet sweeps out equal area.

With **Galileo**'s observations of the heavens with the then newly invented telescope, the overwhelming evidence was for the acceptance of the **heliocentric** model of the universe. However, such was the nature of Galileo's personality that he came into conflict with the powerful catholic church which at that time had not yet incorporated the heliocentric model into religious doctrine. It took another 100 years or so before the heliocentric model gained general acceptance by theologians.

Although Kepler could explain the geometry of planetary motion, he could not explain *why* the planets moved. Kepler proposed that the sun, being at the centre of the universe, emitted "rays" which swept the planets, including the Earth, around the sun. In 1674, Robert **Hooke** proposed that all celestial bodies, (including the sun, planets and the Earth) have an attraction towards their own centres - a **gravitational force**. However, Hooke could not say what the magnitude of this force might be. Isaac **Newton**, in 1666, proposed that the force acting between any two bodies exhibited an **inverse square law**. Newton proposed that this same force extended to planetary bodies and that the force of attraction between the sun and the planets was this same inverse square law. Newton demonstrated that such a law would lead naturally to Kepler's three laws of planetary motion.

5.4.3 Size of the universe

The heliocentric model of the universe required the distance from the Earth to the stars to be a very great distance to account for the apparent lack of motion of the stars as seen from the Earth.

Using a parallax method with the Earth's orbit as the base, **Bessel**, in 1838, measured the parallax of the star 61 Cygni to be 0.30 seconds of an arc which represented a distance of 3 parsecs or 11 light years.

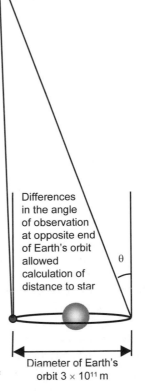

If r = 1.49 × 10¹¹ m is the mean radius of the Earth's orbit around the sun, then the distance d at which the angle θ subtended is one second of an arc is called a parallax second, or **parsec** (pc). 1pc = 3.086 × 10¹⁶ m. It is usually convenient to work in **Megaparsecs**. 1Mpc = 3 × 10²² m. The **light year** is the distance traveled by light in one year. 1 ly = 9.461 × 10¹⁵ m. Therefore 1 pc = 3.261 ly. The **au** is the mean radius of the Earth's orbit and is equal to 1.49 × 10¹¹ m.

Differences in the angle of observation at opposite end of Earth's orbit allowed calculation of distance to star

Diameter of Earth's orbit 3 × 10¹¹ m

Measurements of astronomical distances using parallax are limited by the resolution of the measuring instruments to only the nearest stars (less than about 300 light years), the remainder showing no observable parallax. Further, astronomers of the time were still uncertain about whether **nebulae** were inside or outside the galaxy. In the early part of the 20th century, it was shown using the relative brightness of maximum and minimum in pulsating stars called Cepheid variables that the nebulae which could be resolved into stars were located far outside the Milky Way and were galaxies in their own right. In 1926, Edwin **Hubble**, using the luminosity of **Cepheid variables** within the Andromeda galaxy, estimated that the distance to this galaxy was 720,000 light years, a figure later corrected to 2.36 million light years when the precise nature of the Cepheid variables observed by Hubble were investigated further. Present day techniques can observe objects (quasars) about 6000 million light years away.

5.4.4 Expansion of the universe

In the period 1912 to 1929, astronomers undertook spectral analysis of the light from distant galaxies. It was discovered that as a result of the **Doppler effect**, the positions of spectral lines were displaced towards the red end of the spectrum, implying that the galaxies were moving away from the Earth at incredible speeds. In 1929, **Hubble** found that the shift in the spectral lines, and hence the velocity of recession of the galaxy, was dependent directly on the distance the galaxy was away from the Earth. That is, galaxies are moving away from each other, the further away, the greater their speed.

Hubble's Law

The recession velocity of a galaxy is proportional to its distance away from the point of observation.

The constant of proportionality is called the **Hubble constant** and is estimated to be within 50 and 100 km s^{-1} megaparsec^{-1}

Before 1920, it was generally accepted that the universe was static and unchanging. **Einstein's theory of general relativity** actually predicted an expanding universe so Einstein modified the theory by introducing a **cosmological constant** which had the effect of balancing this expansion exactly with the force of gravity - thus leading to a static universe. In 1922, Alexander **Friedmann** showed that if it were accepted that on a large scale, the universe was of uniform appearance in all directions, then Theory of General Relativity actually correctly predicted an expanding universe - precisely that observed several years later by Hubble.

Although it appears that all galaxies are receding from us, this does not necessarily mean that the Milky Way is at the centre of the universe. The same expansion will be seen anywhere in the universe - just like spots on the surface of a balloon being blown up. Any spot will see the others receding, the further away they are, the faster the recession.

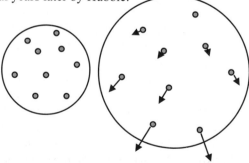

Using the present value of the Hubble constant, and working backwards until "zero" distance, the estimated **age of the universe** is somewhere between 10,000 to 20,000 million years. Other measurements using the age of heavy elements and the age of galaxies put the age of the universe at similar values.

5.4.5 Modern theories of the universe

The synthesis of expansion of the universe and the equations of general relativity by **Friedmann** lead to the conclusion of the existence of a singularity in space and time from which all matter was created. This is called the **Big Bang** theory of the universe. The theory proposes that the universe was created in a gigantic explosion with the creation of Hydrogen and Helium. Space rapidly expanded and the hydrogen and helium cooled as the density decreased and coalesced into stars and galaxies within which the heavier elements were made. Calculations of the age of the universe and the rate of expansion of the universe would result in a residual **background radiation** temperature of about 3 K at our present point in time. This radiation was detected by radio astronomy in 1965, providing good support for the big bang theory.

Big Bang Theory	Steady-State Theory
• Allowed an expanding universe in agreement with Hubble and Friedmann.	• Allowed an expanding universe in agreement with Hubble and Friedmann.
• The Big Bang represents the creation of matter, space and time.	• Allowed an infinite past and an infinite future for the universe (time has no beginning and no end).
• Predicts existence of background radiation pervading the universe which has been experimentally observed.	• The large-scale structure of the universe remains unchanged in accordance with our observations.

In 1948 **Bondi, Gold**, and **Hoyle** presented the **steady-state theory** of the universe as an alternative to the big bang theory. They sought to incorporate the equations of general relativity into a scheme which removed the requirement for a singularity at the beginning of the universe - a feature which could never be opened to investigation and therefore, in their view, objectionable. The key to their scheme was that the universe, on a large scale, is uniform in appearance in both space and time in accordance with Friedmann's model for an expanding universe. The decrease in the density of the universe caused by its expansion is exactly balanced by the continuous creation of matter which formed new galaxies to replace those which had moved on past our field of view. The amount of matter created each year was in the order of 1 particle per km^3 which is within our experimental measure of uncertainty.

5.4.6 Creation of matter

In accordance with the Big Bang theory, it is generally now accepted that all matter, space and time was created from a single point known as a **singularity** in which the universe had zero size and infinite energy.

Quantum theory predicts that matter in the form of a **matter** and **antimatter** can be formed from energy in accordance with the Einstein equation $E = mc^2$. Collisions between matter and antimatter at the big bang produced large amounts of gamma radiation. A slight asymmetry between the amount of matter and antimatter produced an overall net amount of matter in existence.

As the universe expanded, it cooled and elementary particles (**electrons, neutrons** and **quarks**) were able to "condense" to form protons and neutrons and eventually the heavier elements.

Time	Temp °C	Relative Size	Event
0		0	Big Bang
1 sec	10×10^9	10×10^{-10}	Inflation
60 sec	1×10^9	1×10^{-10}	H & He nuclei
1×10^6 yrs	3000	3×10^{-4}	Neutral atoms
2×10^9	5	0.2	Milky Way
5×10^9	1.5	0.66	Solar system
4.6×10^9			Earth
12×10^9	−270	1	Present day

The distribution of matter after the singularity depends on the assumptions made about the mean density of the universe. If the mean density is chosen to be above a critical value, then theory predicts that gravitational forces will eventually overcome those from expansion of the singularity and the universe will collapse (**the Big Crunch**) - a closed universe. If the mean density is below a critical value, then the expansion of the universe will continue indefinitely - an open universe in which eventually all stars will die out - the **Big Chill**.

When the mean density of the universe is calculated, by first estimating the mass of a galaxy by observing the motion of its stars and multiplying the mass of each galaxy by the number of galaxies, the mean density of the universe is found to be very much below the critical value. This discrepancy means that there appears to be a of substantial amount of invisible or **dark matter** within the universe that is at the present time, unaccounted for.

5.4.7 Formation of stars and galaxies

Galaxies

Three minutes after the big bang, all the matter in the universe consisted of about 75% **hydrogen** and 25% **helium** nuclei. As the universe cooled, these nuclei began to capture electrons creating neutral atoms. Gravitational forces allowed these neutral atoms to coalesce into gas clouds. Gravitational forces from other parts of the universe caused some gas clouds to rotate and with continued collapse, the loss of gravitational potential energy resulted in a faster rate of rotation resulting in disk or spiral type galaxies such as the **Milky Way**.

At the formation of the Milky Way, the only elements were hydrogen and helium. As the oldest stars experienced supernova explosions, the heavier elements were created and dispersed throughout the galaxy.

Stars

With increasing density of material within a forming galaxy due to gravitational attraction, parts of the gas cloud begin to rotate on their own. Increasing collisions between molecules as they coalesce result in an increase in temperature until the conditions for nuclear fusion are reached. The resulting heat generated by the fusion reaction gives rise to an increase in pressure which is balanced by gravitational collapse resulting in a stable system called **stars**.

Milky Way

The Milky Way contains about 100×10^9 stars and has the form of a flat disk rotating about its centre. The size of the Milky Way is about 30,000 parsecs (100,000 light-years) in diameter. The sun takes approximately 250 million years to travel once around the center of the Milky Way.

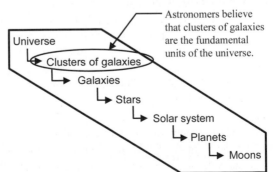

Astronomers believe that clusters of galaxies are the fundamental units of the universe.

5.4.8 Luminosity

The **intensity** or **brightness**, I, of the light emitted by a star is what the eye responds to and is a measure of **radiant flux density** in W m^{-2}. The **intrinsic luminosity**, L, of a star is a measure of the star's **radiant power** output (in Watts) and this is of significant interest to astronomers.

The **apparent brightness** (Wm^{-2}) of a star (i.e. the brightness we see on Earth) depends on the star's intrinsic luminosity (or radiant power output in Watts) *and* the distance at which the star is from us. It can be conveniently measured using a photometer.

$$I = \frac{L}{4\pi d^2}$$

If the distance d to a star is known, (by parallax measurements) and the apparent brightness I measured (with a photometer) then the **intrinsic luminosity** L of the star be calculated using the inverse square law.

The intrinsic luminosity or radiant power (W) depends on the size of the star and its intrinsic brightness or intensity (W m^{-2}) at its surface. A large star with low intrinsic brightness (i.e. a low temperature) may have the same intrinsic luminosity (power output) as a small star of large intrinsic brightness.

The brightness that a star would have if it were at a distance of exactly 10 parsecs away is called (by definition) the **absolute brightness** or **absolute magnitude** M and allows us to compare the brightness of stars independent of their distance from us.

The absolute magnitude is conveniently expressed in terms of the ratio of the intrinsic luminosity of the star and that of the sun. Mathematically, the **absolute magnitude** of a star is calculated from:

$$M = 4.75 - 2.5\log\frac{L}{L_o}$$

The temperature of the sun can be obtained from spectroscopic measurements. The intrinsic luminosity is then given by the Stefan-Boltzmann radiation emission law using the known radius of the sun. At R = 0.7 × 10^9 m T = 5800K, L$_o$ = 3.90 × 10^{26} W.

If the emission spectra of two stars are compared and show the same shape (i.e. same colour) then we can say that they are at the same temperature. This means that any difference in their total power output is due to a difference in their size.

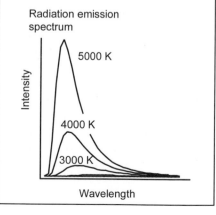

Radiation emission spectrum

5.4.9 Lifecycle of stars

In its early stages a **protostar** star is formed by coalescence of the gas cloud which has the appearance of a disk. **Protoplanets** may also begin to form. Formation of a hot core takes about 1×10^6 years. Approximately 70 $\times 10^6$ years later, contraction causes a rise in temperature which is enough to initiate nuclear fusion and the star is then on the **main sequence**.

During the **red giant** stage, the surface temperature is about 3000 K. There are periods of instability at this point where the outer atmosphere of the star is blown off as a **planetary nebulae** by the pressure from below. About 75000 years later, the core of the star collapses to form a very faint **white dwarf**. Gravitational collapse in a white dwarf is countered by quantum effects acting on their electrons within it. If the star is more massive than the sun, it moves along through the red giant stage more quickly and instead of forming a white dwarf the inner portions collapse to form a **neutron star** or alternately a **black hole**. The shock wave from this causes the outer layers to explode as a **supernova**.

Hertzsprung-Russell diagram

In **main sequence** stars, of which the sun is one, the source of energy is the nuclear binding energy associated with the fusion of hydrogen into helium. In the **Hertzpsrung-Russell** diagram, for main sequence stars, the larger stars are at the top left while the lower mass stars are at the lower right. For stars with a size similar to the sun, the main sequence lasts for about 1×10^{10} years. As the amount of hydrogen becomes depleted, gravitational forces collapse the star inwards until the temperature is high enough to cause the fusion of helium into heavier elements. The outer layers of the star expand outwards and the star enters the **red giant** branch.

5.4.10 Energy and matter

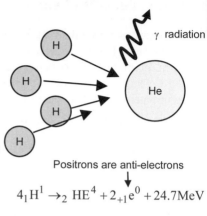

γ radiation

Nuclear fusion in main sequence stars results in the formation of a helium nucleus from four protons. This does not happen in one step, but through a series of reactions. The energy (24.7 MeV) is released in the form of **gamma rays** and kinetic energy of the helium nuclei. **Positrons** created during the process are annihilated by combining with free electrons to produce 2 MeV of energy in the form of gamma rays. The total energy released is 26.7 MeV.

Positrons are anti-electrons

$$4_1 H^1 \rightarrow_2 HE^4 + 2_{+1} e^0 + 24.7 MeV$$

Very large stars (approximately 5 times the mass of the sun) go through their evolutionary stages very quickly compared to the Sun. As the supply of hydrogen in such a star becomes depleted, the star contracts until the temperature becomes high enough to allow the fusion of helium, which is now in abundance, into heavier elements such as carbon and nitrogen. The outer atmosphere of the star increases and it becomes a **red giant**. Nuclear fusion proceeds providing radiation pressure outwards to counteract the inward pull of gravity and elements up to the atomic mass of iron are created. Then, the star's core starts to contract rapidly since once iron is formed, fusion cannot proceed any further.

The sudden collapse of the core releases a vast amount of gravitational potential energy which blows away all the outer parts of the star in a violent explosion called a **supernova**. All the elements with atomic weights greater than iron are formed during this explosion.

In the central core, protons and electrons are formed into neutrons and the core becomes a neutron star. The **gravitational collapse** also causes the core to spin rapidly and the charged particles in the vicinity emit radiation as a result which is detected by us as regular pulses and these objects are thus known as **pulsars**. Alternatively, the gravitational collapse is so large that a black hole is formed.

The **Crab nebula** is the remains of a supernova explosion in our galaxy which was observed first hand by astronomers in 1054.

5.4.11 Solar system

During the formation of a star, heavier molecules, consisting metals and water, may form what is called an **accretion disk**. The molecules in the accretion disk cool and sometimes collide with each other and form large dust particles. These dust particles have their own gravitational field which may attract more dust particles. The process rapidly accelerates and **protoplanets** are formed.

As the planets begin to form, the evolving sun begins to radiate both energy in the form of electromagnetic radiation and also charged particles called the **solar wind**.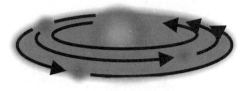

The solar wind pushes any remaining gas molecules and dust out of the solar system leaving only the planets behind.

Larger planets have a gravity large enough to retain some of the original gas molecules and form **gas giants**. Others form smaller **rocky planets** initially without an atmosphere. Inside the planets, heavy (more dense) elements gravitate towards the centre leaving the lighter elements near the surface.

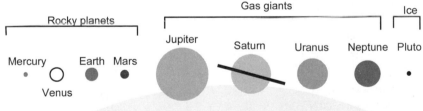

Planet	Mass x10²⁴kg	Diameter km	Orbit radius 10⁹m	Orbital eccentricity	Period
Mercury	0.328	4800	57.7	0.206	88 days
Venus	4.83	12310	107.7	0.007	224.3 days
Earth	5.98	12757	149	0.017	365.3 days
Mars	0.637	6790	226.9	0.093	687.1 days
Jupiter	1900	142850	774.8	0.048	11.9 years
Saturn	567	115000	1415.5	0.056	29.6 years
Uranus	88	47150	2859.3	0.047	84.0 years
Neptune	103	44700	4480.3	0.009	164.8 years
Pluto	1.1	6000	5888.5	0.249	248.4 years

5.4.12 Sun

Technically speaking, the **sun** is a gaseous body composed of hydrogen (75%) and helium (25%) with very small amounts of oxygen and other elements. The surface temperature of the sun is about 5800 °C. The interior temperature is very much higher than this ($\approx$ 15,000,000 °C).

Within this very hot interior, hydrogen nuclei (which are just single protons) undergo nuclear fusion to form helium nuclei. The actual fusion reaction is complex, but overall, four hydrogen nuclei combine to form one helium nucleus. The mass of a helium nucleus is less than the original four hydrogen nuclei, and this difference in mass is converted to energy

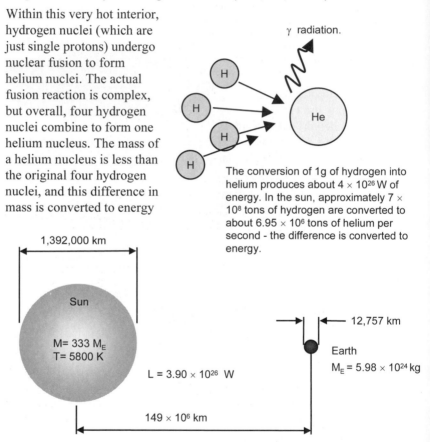

γ radiation.

H

H

H

H

He

The conversion of 1g of hydrogen into helium produces about 4×10^{26} W of energy. In the sun, approximately 7×10^8 tons of hydrogen are converted to about 6.95×10^6 tons of helium per second - the difference is converted to energy.

1,392,000 km

Sun

M= 333 M_E
T= 5800 K

L = 3.90×10^{26} W

12,757 km

Earth
$M_E = 5.98 \times 10^{24}$ kg

149×10^6 km

The sun is a medium mass star presently within its **main sequence**. It is estimated to be about 4500 million years old, and it will be at least another 2000 million years before the hydrogen fuel supply begins to run out. The mass of the sun is estimated to be approximately 2×10^{30} kg.

5.4.13 Solar atmosphere

Sunspots are seen as dark spots on the Sun's surface (first seen by Galileo in 1610). They appear dark by contrast with the brighter and hotter surrounding surface. They consist of matter welling up from the interior of the sun being affected by the sun's magnetic field. They can last a few hours or several months and can occur in groups on on their own. They usually occur in waves going through a maximum number every 11 years or so. Galileo observed the rotation of sunspots across the face of the sun and concluded that the sun is rotating once every 26 days.

Solar prominence consist of atoms and ions ejected from sunspots and pass through the chromosphere

Photosphere is the disk of the sun

Limb darkening is evidence of the gaseous nature of the sun

Chromosphere is cooler than the photosphere.

Solar flares are usually associated with sunspots and faculae. They give rise to bursts of intense ultra violet radiation which cause disturbances in the ionosphere of the Earth resulting in disruption of telecommunications. Electrically charged particles are also ejected which interact with the Earth's magnetic field and causing the aurora borealis. This constant stream of electrically charged particles from the sun is called the **solar wind**.

Faculae are very bright areas usually near sunspots and can be seen near the limb.

The **Corona** is a tenuous atmosphere of molecules, atoms, ions. Can usually only be seen during a total eclipse and extends to about 25,000,000 km from the sun.

In the period 1826 and 1843, Heinrich Schwabe discovered the cycle of sunspot activity in the sun. At the beginning of each cycle, sunspots occur in high latitudes on the sun and reach a maximum number within about 4 years. They then occur at lower and lower latitudes and sometimes die off all together within about 7 years of reaching the maximum number.

5.4.14 Solar wind

Sunspots look dark because they are at a lower temperature than the surrounding regions. In fact, the temperature of sunspots is about 4,000 K and can be up to 50,000 km in diameter. In addition to electromagnetic radiation, the Sun also emits a stream of charged particles which we call the **solar wind**. These particles travel with a velocity of about 450 km s^{-1}. The solar wind consists of protons, electrons and helium nuclei. The solar wind is a different thing to **cosmic rays**. Cosmic rays arrive at the Earth from *all directions* and are thought to consist mainly of protons.

The solar wind, consisting of charged particles in motion, create disturbances in the Earth's magnetic field. These disturbances are particularly pronounced during sunspot activity. Due to the shape of the Earth's magnetic field, charged particles tend to twist in spirals about the Earth's magnetic lines of force and are drawn to the magnetic poles where they can travel downwards and cause ionisation of air molecules which result in the **aurora** displays.

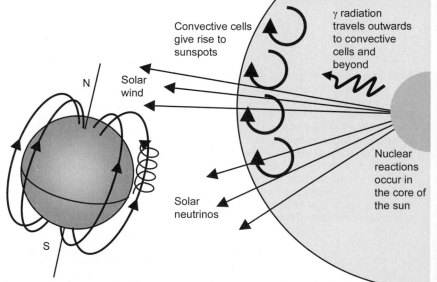

Convective cells give rise to sunspots

γ radiation travels outwards to convective cells and beyond

N

Solar wind

Solar neutrinos

Nuclear reactions occur in the core of the sun

S

5.4.15 Solar radiation

As **gamma rays** travel outwards from the centre of the sun, their energy is continuously absorbed and re-emitted at lower wavelengths. By the time this radiation reaches the surface of the sun, it is mostly visible light.

The sun can be regarded as a **black body**. That is, it emits the maximum amount of radiation possible over all wavelengths. By the **Stefan-Boltzmann law**, the rate of radiant energy emission is given by:

$$\dot{Q} = \sigma \varepsilon A T^4$$

σ = 56.7 × 10⁻¹² kW m⁻² K⁻⁴
Stefan-Boltzmann constant
ε = 1 for a black body.

Assuming the surface area of the sun is $4\pi r^2$ where R = 0.7 × 10⁹ m, then, setting T = 5800K we have:

$$\dot{Q} = 56.7 \times 10^{-12} (1) 4\pi \left(0.7 \times 10^9\right)^2 5800^4$$
$$= 3.9 \times 10^{26} \, W$$

If the radius of the Earth's orbit is 1.49 × 10¹¹ m, then the intensity of radiation received just outside the atmosphere is:

$$I = \frac{3.9 \times 10^{26}}{4\pi \left(1.49 \times 10^{11}\right)^2}$$
$$= 1.4 \text{kWm}^{-2}$$

Not all of the radiation from the sun strikes the Earth's surface. That which is not reflected back into space or absorbed in the atmosphere is mostly in the visible and radio parts of the electromagnetic spectrum. Ultraviolet, X ray and gamma ray radiation is filtered out by the **atmosphere**.

If a **solar collector** is placed out in the direct sun on the Earth's surface, the plate receives both direct and diffuse radiant energy. If say 40% is lost due to atmospheric absorption, and an extra 20% is obtained by diffuse radiation, then the radiation intensity at the Earth's surface is approximately 1 kW m⁻².

Not all frequencies of radiation from the nuclear reactions within the core of the sun are emitted from the sun due to **absorption** within the sun itself. Wavelengths of light in the absorption spectrum of hydrogen and helium and other elements appear as "dark" lines in the overall emission spectrum of the sun. One set of absorption lines could not be identified in 1815 when these spectra were first studied prompting the announcement of the discovery of a new element **helium** - the "sun element".

5.4.16 Earth

Mass	5.98×10^{24} kg
Radius	6.38×10^6 m
Density	5520 kg m^{-3}
Escape velocity	11200 m s^{-1}
Distance from Sun	1.49×10^{11} m
Rotation period	23.93 hours
Length of year	365.26 days
Tilt of axis	23.4°
Orbit eccentricity	0.017
Age	4.5 billion years

The core of the **Earth** is thought to be solid iron or iron and nickel - materials similar to that found in meteorites which fall on the Earth from space. The outer core is thought to be liquid since transverse **S waves** from **earthquakes** do not appear to travel through it. The inner core is thought to be solid.

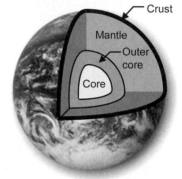

The **mantle** is solid rock mainly consisting of the minerals olivine and pyroxene. Local melting occurs within the mantle which sometimes erupts as **volcanoes** at the Earth's surface

The **crust**, although very thin, can be divided into two zones, the lower of which is about 3–5 km thick and consists of basalt. The upper zone is the continental crust having varying composition mainly comprising granite.

Layer	Depth (km)
Crust	0 – 40
Mantle	40 – 2700
Core	2700 – 5200
Inner Core	5200 – 6371

The underlying cause of activity within the crust (the formation of mountains, volcanoes, earthquakes etc) are thought to be a result of movement of plates which are in constant motion. The plates are in contact with each other at margins and their movement creates upwelling of basalt from the mantle along constructive margins (usually on the sea floor) and downwards motion of basalt and piling up of the upper crust in destructive margins. This movement gives rise to **continental drift**.

5.4.17 Earth's magnetic field

In 1600, William **Gilbert**, court physician to Queen Elizabeth, proposed that the Earth is a magnet and that a freely suspended magnet on the Earth's surface would line up approximately in a north south direction.

We now know that the Earth's magnetic field arises due to electric currents in the core and its shape is similar to that which would result if a short bar magnet were placed at the centre of the Earth. The direction of the **magnetic poles** (where the magnetic lines of force are vertical) are about 15° from the Earth's rotational axis. The magnitude of the field at the Earth's surface is approximately 10^{-4} T (1 **Gauss**).

The magnetic lines of force are not parallel with the surface of the Earth. The angle which the lines of force make with the surface is called the **inclination**. At the magnetic poles the inclination is 90°. At the magnetic equator, the inclination is zero.

The magnetic field of the Earth extends outwards for about 5 times the radius of the Earth. At larger distances, the field is distorted by the charged particles of the solar wind.

Solar wind

The component of velocity which is perpendicular to the field lines causes the particles to travel in a circle. The component of velocity perpendicular to the field lines causes positive particles to travel towards the north magnetic pole. The overall path is a spiral.

During periods of high sunspot activity, charged particles from the sun (the **solar wind**) interact with the Earth's magnetic field and travel in a spiraling path according to the right hand rule. At regions of high magnetic field strength, such as at the poles, they are mostly reflected back along the field lines and are thus trapped in what are called the **Van Allen** radiation belts thus shielding us from potentially ionizing radiation.. Since at the poles charged particles are drawn in towards the Earth's surface, they ionize air molecules causing the **aurora** displays.

5.4.18 Atmosphere

The **atmosphere** is a relatively thin layer of gas which covers the Earth. 80% by mass of the atmosphere exists in the lowest 12 km of the atmosphere which extends some 100 km into space.

Ozone (O_3) is produced in the stratosphere by ultraviolet light acting on oxygen molecules causing them to split and recombine into O_3. Ozone is very reactive and if it were not continually regenerated, would disappear within a few months. The continued absorption of ultraviolet radiation in the **stratosphere** to produce ozone shields us from its harmful effects. Degeneration of the ozone layer by chloroflourocarbon (**CFC**) gases is a major concern. Approximately 34% of the incident radiation on the Earth is reflected back into space by clouds (called the **albedo effect**). 19% is absorbed by the atmosphere, and 47% is absorbed at the Earth's surface.

Water vapour, carbon dioxide, methane, nitrous oxides, and CFC's are transparent to visible electromagnetic radiation but absorb infrared radiation being emitted from the Earth's surface. The presence of these gases in the atmosphere causes the surface of the Earth to be warmer than it otherwise would be. To some extent, this **greenhouse effect** is a natural phenomenon. However, human activity during the past 100–200 years has increased the level of greenhouse gases to the extent that **global warming** is of significant concern.

Composition of the atmosphere	
	%Vol
Nitrogen	78.09
Oxygen	20.95
Argon	0.93
Carbon Dioxide	0.023-0.050
Neon	0.0018
Helium	0.0005
Krypton	0.0001
Hydrogen	0.00005
Xenon	0.000008
Methane	0.00017
Nitrous Oxide	0.000003
Carbon Monoxide	0.000005
& trace amounts of Sulfur Dioxide, Nitrogen Dioxide, Ammonia and Ozone	

The **ionosphere** is created by ionisation of air molecules by radiation from the sun. Both ionised atoms and free electrons are produced, both of which serve to reflect radio waves of certain frequencies allowing long distance radio communications.

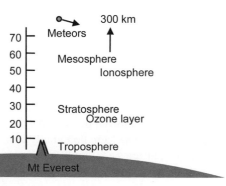

5.4.19 Examples

1. The planet Pluto orbits the sun at a mean distance of 5.9×10^{12}m. Determine the period of Pluto's orbit T_p if the Earth takes one year to orbit the sun at a radius of 1.49×10^{11} m.

Solution:

$$\frac{1^2}{\left(1.49 \times 10^{11}\right)^3} = \frac{T_p^{\,2}}{\left(5.9 \times 10^{12}\right)^3}$$

$$T_p = 249 \, yr$$

Kepler's third law states that the square of the period of the orbit of a planet is proportional to the cube of planet's distance from the sun, so we can use information about the Earth's orbit to calculate the period of any other planet if the radius of the orbit is known.

2. Calculate the mass of the Earth given the radius of the Earth to be 6.38×10^6 m and acceleration due to gravity as 9.81 m s^{-2}.

Solution:

$$F = G \frac{m_1 m_E}{R_E^{\,2}} = m_1 g$$

$$m_E = \frac{g R_E^{\,2}}{G}$$

$$= \frac{9.8 \left(6.38 \times 10^6\right)^2}{6.673 \times 10^{-11}}$$

$$= 5.98 \times 10^{24} \, kg$$

This calculation was first done by Cavendish using a value of G that he measured between two known masses in his laboratory.

Materials data

α_{Al}	$= 7.2 \times 10^{-5}\,^{\circ}C^{-1}$	M_{air}	$= 0.02892$ kg mol^{-1}
α_{steel}	$= 1.2 \times 10^{-5}\,^{\circ}C^{-1}$	M_{O2}	$= 0.032$ kg mol^{-1}
$\alpha_{concrete}$	$= 10 \times 10^{-5}\,^{\circ}C^{-1}$	M_H	$= 0.001008$ kg mol^{-1}
α_{brass}	$= 20 \times 10^{-6}\,^{\circ}C^{-1}$	c_{pAir}	$= 1.005$ kJ kg^{-1} K^{-1}
α_{Cu}	$= 16.5 \times 10^{-6}\,^{\circ}C^{-1}$	c_{vAir}	$= 0.718$ kJ kg^{-1} K^{-1}
α_{glass}	$= 7.75 \times 10^{-6}\,^{\circ}C^{-1}$	γ_{air}	$= 1.4$
ρ_{oil}	$= 915$ kg m^{-3}	P_{atm}	$= 101.3$ kPa
ρ_{Fe}	$= 7800$ kg m^{-3}	I_o	$= 1 \times 10^{-12}$ W m^{-2}
ρ_{water}	$= 1000$ kg m^{-3}	ρ_{Ag}	$= 1.64 \times 10^{-8}\,\Omega$ m
ρ_{air}	$= 1.2$ kg m^{-3}	ρ_{Cu}	$= 1.72 \times 10^{-8}\,\Omega$ m
ρ_{Hg}	$= 13600$ kg m^{-3}	ρ_{Al}	$= 2.83 \times 10^{-8}\,\Omega$ m
ρ_{Pb}	$= 11300$ kg m^{-3}	ρ_W	$= 5.5 \times 10^{-8}\,\Omega$ m
$\rho_{glycerine}$	$= 1300$ kg m^{-3}	E_{steel}	$= 210$ GPa
ρ_{steel}	$= 7700$ kg m^{-3}	E_{Al}	$= 70$ GPa
c_{water}	$= 4186$ J kg^{-1} K^{-1}	E_{glass}	$= 70$ GPa
c_{Fe}	$= 540$ J kg^{-1} K^{-1}	E_{wood}	$= 14$ GPa
c_{Al}	$= 920$ J kg^{-1} K^{-1}	E_{Fe}	$= 1.9 \times 10^{11}$ Pa
c_{steel}	$= 450$ J kg^{-1} K^{-1}	E_{brass}	$= 100$ GPa
c_{ice}	$= 2300$ J kg^{-1} K^{-1}	E_{Cu}	$= 1.1 \times 10^{11}$ Pa
c_{oil}	$= 2100$ J kg^{-1} K^{-1}	B_{Al}	$= 7.46 \times 10^{10}$ Pa
L_{Vwater}	$= 22.56 \times 10^5$ J kg^{-1}	B_{water}	$= 2.04 \times 10^9$ Pa
L_{Fwater}	$= 3.33 \times 10^5$ J kg^{-1}	χ_{O2}	$= 1.9 \times 10^{-6}$
k_{steel}	$= 50.2$ W m^{-1} K^{-1}	χ_{Al}	$= 2.2 \times 10^{-5}$
k_{Al}	$= 205$ W m^{-1} K^{-1}	χ_{Pt}	$= 2.6 \times 10^{-4}$
k_{Cu}	$= 385$ W m^{-1} K^{-1}	m_E	$= 5.98 \times 10^{24}$ kg
$k_{plaster}$	$= 0.12$ W m^{-1} K^{-1}	R_E	$= 6.38 \times 10^6$ m
$k_{fibreglass}$	$= 0.039$ W m^{-1} K^{-1}	R_{MOrbit}	$= 3.81 \times 10^8$ m
k_{air}	$= 0.034$ W m^{-1} K^{-1}	R_{EOrbit}	$= 1.49 \times 10^{11}$ m
k_{brick}	$= 1.33$ W m^{-1} K^{-1}	$\mu_{r\ water}$	$= 0.999991$
k_{glass}	$= 0.79$ W m^{-1} K^{-1}	$\varepsilon_{r\ water}$	$= 81$
k_{water}	$= 0.65$ W m^{-1} K^{-1}	$\mu_{r\ air}$	$= 1.00000036$
σ_{Ag}	$= 6.2 \times 10^7$ S m^{-1}	$\varepsilon_{r\ air}$	$= 1.0006$
σ_{Al}	$= 3.7 \times 10^7$ S m^{-1}	$\varepsilon_{r\ glass}$	$= 6$
σ_{water}	$= 1 \times 10^{-3}$ S m^{-1}		
$\sigma_{salt\ water}$	$= 4$ S m^{-1}		
σ_{glass}	$= 1 \times 10^{-12}$ S m^{-1}		

Index

Useful information

Speed of light	c	2.99793×10^8 m s^{-1}
Charge on an electron	q_e	1.6022×10^{-19} C
Boltzmann's constant	k	1.38062×10^{-23} J K^{-1}
Stefan-Boltzmann constant	s	5.6697×10^{-8} W m^{-2} K^{-4}
Planck's constant	h	6.626×10^{-34} J s
Rydberg constant	R	1.0973731×10^7 m^{-1}
Avogadro's number	N_A	6.02217×10^{23}
Universal gas constant	R	8.3143 J mol^{-1} K^{-1}
Mass of an electron	m_e	9.1096×10^{-31} kg
Mass of a proton	m_p	1.673×10^{-27} kg
Mass of a neutron	m_n	1.675×10^{-27} kg
Atomic mass unit	amu	1.6602×10^{-27} kg
Coulomb force constant	k	9×10^9 m^2 C^{-2}
Permittivity of free space	ε_o	8.85×10^{-12} F m^{-1}
Permeability of free space	μ_o	$4\pi \times 10^{-7}$ Wb A^{-1} m^{-1}
Acceleration due to gravity	g	9.81 m s^{-2}
Standard atmospheric pressure	p_{atm}	101.325 kPa
Absolute zero		-273.15°C
Gravitational constant	G	6.673×10^{-11} N m^2 kg^{-2}
Electron volt	eV	1.6022×10^{-19} J
Mass of the earth	m_E	5.983×10^{24} kg
Radius of the earth	r_E	6.371×10^6 m
Mass of the sun	m_S	1.999×10^{30} kg

One **atomic mass unit** (amu) is 1/12th the mass of a carbon 12 atom by international agreement. The atomic weight of C_{12} is 12 amu and 1 amu is 1.6602×10^{-27} kg.

The total sum of the atomic weights for a molecule of substance is called the **molecular weight** m.w. The molecular weight of water, H_2O, is 16 + 2 = 18 amu.

The molecular weight expressed in grams contains one mole of molecules (or more precisely, one gram-mole). 12 grams of C_{12} contains 6.02×10^{23} atoms. 18 grams of H_2O also contains 6.02×10^{23} molecules. This is Avogadro's number.

The molar mass M_m is the mass of one mole of molecules (or atoms) expressed in kilograms. The molar mass of H_2O is 0.018 kg.